DATE			

MACHINE DYNAMICS

MACHINE DYNAMICS

HENRY J. SNECK
Rensselaer Polytechnic Institute

PRENTICE HALL, Englewood Cliffs, New Jersey 07632

Sneck, H. J. (Henry J.)
    Machine dynamics / by H.J. Sneck.
        p.   cm.
    ISBN 0-13-543299-5
    1. Machinery, Dynamics of.   I. Title.
TJ170.S54   1991
621.8'11--dc20   90-6738

Editorial/production supervision
    and interior design: *Mary Kathryn Leclercq*
Cover design: *Joe DiDomenico*
Prepress buyer: *Linda Behrens*
Manufacturing buyer: *David Dickey*

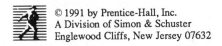
Printed in the United States of America

10 9 8 7 6 5 4 3 2 1

ISBN   0-13-543299-5

Prentice-Hall International (UK) Limited, *London*
Prentice-Hall of Australia Pty. Limited, *Sydney*
Prentice-Hall Canada Inc., *Toronto*
Prentice-Hall Hispanoamericana, S.A., *Mexico*
Prentice-Hall of India Private Limited, *New Delhi*
Prentice-Hall of Japan, Inc., *Tokyo*
Simon & Schuster Asia Pte. Ltd., *Singapore*
Editora Prentice-Hall do Brasil, Ltda., *Rio de Janeiro*

CONTENTS

PREFACE

At some time during a mechanical engineering student's academic program, he or she is required to take a first course or courses in mechanics. The content of these courses is usually divided in two parts, statics and dynamics. Because a knowledge of mechanics is fundamental to the understanding of so many other areas of engineering, it is customary to give these courses in the early years of the student's education. Typically, these courses immediately follow the mathematics courses containing differential calculus, integral calculus, and vector algebra.

Unfortunately, the student is poorly equipped at this time to appreciate, or even to cope with, the engineering applications of mechanics. This is especially true of dynamics, which in many texts is slighted in favor of the easier-to-teach statics material. What little dynamics is done usually consists of examining problems unrelated to realistic engineering applications. Considering the student's background at this stage of his or her education, one cannot hope to do much more than examine cylinders rolling down an inclined plane or a gyroscope rotating about a fixed point. As a result, most students come away from such courses with the uneasy impression that mechanics must have some relevance to the solution of real engineering problems, but they are not sure just how.

Acknowledging this situation, most mechanical engineering curricula require a second mechanics-based course to be taken near the end of the student's program.

The purpose of this course is to tie the formalized mechanics concepts to more realistic engineering situations at a point in the student's education when he or she is better prepared to deal with the engineering aspects of mechanics and is less concerned with grasping the fundamental concepts. That is not to say that it can be assumed that the student really understands mechanics by just successfully completing a course in mechanics. Anyone who has taught a second, follow-on course in mechanics will attest to that. Thus, a second-time-around mechanics-based course can serve two purposes: to teach engineering applications, but secondarily, and just as important, to reinforce and further develop the skills taught in the earlier mechanics course.

The question is how to do this, since a rehash of the standard first mechanics course is clearly out of order. That decision is based on two considerations: what needs to be reinforced and what can be added. Since kinematics is the heart of dynamics, it is a leading candidate for reinforcement. The "modern" vector methods used in most mechanics texts are clearly the techniques of choice for elegance and efficiency. However, from the student's point of view, they are quite sterile, often confusing, and even border on magic, especially when moving reference frames are used. Mechanics texts seldom point out that moving reference frames are a convenience that permit the orderly construction of the kinematics. Never mentioned is that they are often a necessity in view of the way experimental or monitoring data are obtained. Little wonder students are confused about relative motions, Coriolis accelerations, angular velocities and accelerations, etc. It is difficult to develop a physical feel for these quantities when they are enmeshed in the formalization of vector algebra. It is in this regard that graphical techniques have much to offer the student. It is through the graphical analysis of planar motions (by far the most common in machine dynamics) that the student can, perhaps for the first time, actually see the various components of the velocity and acceleration vectors. Not only are graphical techniques quick and easy to apply, they also pictorially illustrate the direction and magnitude of each component. The relative contribution of each component is obvious to the designer at a glance and so is the effect of changing input variables.

To the student and instructor alike, graphical kinematics may at first seem archaic, a throwback to the prevector era. This is not true. It is simply an application of the fundamental concept of a vector, i.e., a directed line segment, without the use of base vectors. All of the same conceptual content is still there, the same thought processes must be executed, and the same physics and geometry still apply. As an added bonus, one finds that the kinematic matching conditions at interfaces are actually easier to satisfy graphically than when using base vectors. And as if all these advantages were not enough to recommend the method, we are now seeing, with the aid of the high-speed computer, a revival of graphic kinematic presentations.

Thus, the rationale is given for a chapter devoted mainly to graphical kinematics (Chapter 2). Analytical methods are not altogether neglected, however, with the introduction of complex-variable methods!

Having commenced with a graphical technique, it would seem logical to continue to use it in dynamic analyses. Not only is that feasible, it is almost a necessity considering the mathematical alternatives. The analytical tools for the dynamic analysis of interconnected bodies are certainly available, but clearly beyond the scope of a machine dynamics text aimed at the undergraduate engineering student. Except for the simplest systems, these more advanced techniques are best executed via computer codes, which, for the most part, takes one out of the classroom arena. Hence, the dynamics chapter (Chapter 3) follows the kinematics chapter, continuing the graphical approach onto planar dynamics of rigid machine elements.

The kinematics and dynamics chapters are basically a review of the rigid-body mechanics principles taught in the basic mechanics course couched in the framework of a graphical representation. The equations are the same, the technique different, and we hope the understanding broadened. The next two chapters take an entirely different path, one that is almost certainly new to the students. The method is essentially an energy-based one that introduces the student to the Lagrange equations (Chapter 4). In the past, the Lagrange method has been reserved for graduate analytical dynamics courses, where they are developed and employed in their most general sense. The mathematical sophistication required to use the equations at this level is probably beyond the undergraduate's ability. However, in the context of most engineering applications, the use of these equations is well within the capabilities of the undergraduate by learning to follow a few basic rules. Experience has shown that with a little help, the student can grasp the formalities of the technique rather quickly. The chapter on the Lagrange equations introduces them using simple mechanics problems rather than machine dynamics applications. Since this is "new mechanics," the machinelike applications are reserved for the following chapter.

Actually, the chapter on Lagrange equations has little value in its own right. It only illustrates that certain mechanics problems are quicker and easier to set up using the Lagrange equations than by the more commonly used free-body method. If that were the only motivation for introducing the Lagrange equations, there would be little justification for their introduction. The real benefits of this chapter are realized in the following chapter, which, by its heading, indicates that it will consider a variety of machine dynamic applications. Although many of the examples in this chapter (Chapter 5) deal with vibrations, the chapter is not a typical vibrations presentation. The applications considered are restricted to vibrational problems commonly encountered in machinery, their causes, and their cures. Some care in the selection and ordering of the topics has been exercised so that the student is led logically from one level of complexity to the next.

The Lagrange equations provide the vehicle for the development of the mathematical models of the machine. The economies that they provide allow the analysis to concentrate on the results and not on the derivation, something that would not be possible if each situation were to be attacked using free-body equilibrium techniques. Because the derivations of the model equations are deemphasized, the presentation can focus on the machine dynamic consequences of such things as foundation stiffness, shaft unbalance, modal coupling, etc. Although not exhaustive, the treatment of machinery dynamics is sufficiently broad to at least introduce the student to the problems most often encountered by the designer.

A comment on the style of this chapter. It differs from the previous three in that it covers a number of topics that are progressively linked, one to the other, as one proceeds through the chapter. Each topic deals, however, with a specific machine dynamic problem. Unlike the previous chapters, where it is relatively easy for the instructor to devise similar illustrative problems for class presentation and discussion, the specificity of each topic makes that much more difficult. Candidates for class presentation could be drawn from the accompanying problem set for the chapter. Some of these problems require considerable sophistication and might be appropriate homework assignments if they were first started in the classroom as examples.

The author is indebted to his colleague Professor H. A. Scarton for his helpful suggestions about the text material and the numerous exercise problems that he contributed. Support and encouragement came from Professor F. F. Ling, Chairman of the Mechanical Engineering Department at Rensselaer Polytechnic Institute. Betty Alix and Geri Frank deserve special credit for the text typing, and my wife Erye for her patience throughout.

Henry J. Sneck

MACHINE DYNAMICS

1

INTRODUCTION

Machines consist of interconnected moving and stationary elements. These elements are designed to perform certain desired motions. Those motions are accompanied by a transfer of power. Machine dynamics, as the name implies, is concerned with the forces, moments, and power transfers that result from or create the element motions.

The dynamic analyses of machines are, of course, based on the laws of Newtonian mechanics. When constructing a mathematical model of a machine, it is convenient, but not always possible, to lump the elements into rigid masses or massless elastics, according to their relative stiffnesses. If this results in a reasonable representation, the equations of motion are ordinary rather than partial differential equations. In principle, these equations could be applied to every lumped element of the machine using free-body methods. When there are a large number of components, this approach results in a large number of simultaneous equations. The solution of these equations would yield, in overwhelming detail, the behavior of the machine at every instant of its operation. In the design of most machines, this approach is neither economically feasible nor technically desirable. Common sense and experience will usually indicate which parts of the machine are most likely to cause trouble. For example, moving elements with small cross-sectional areas or abrupt changes in area are likely to develop repetitive stresses that can lead to fatigue failure. Bearing loads at the juncture of connected elements can be excessive to the point where seizure can occur. Unwanted motions such as shaft whirl or foundation flexure may arise to jeopardize the safe operation of the machine.

The experienced designer knows where to anticipate problems before the machine is constructed and will take measures to avoid these problems. Should unforeseen problems arise once the machine is built, the designer must be prepared to diagnose and cure the ailment. In either case, the solution is usually a compromise between opposing tendencies. The choice of whether to go bigger or smaller, faster or slower, heavier or lighter, etc. hinges on finding the safe middle ground between conflicting trends. Knowledge of the techniques that quantify the behavior of the elements involved is essential if these choices are to be made intelligently. No text can possibly anticipate in detail every possible problem that can arise. There are, however, certain general techniques, mathematical and graphical, that permit the designer to isolate and analyze most of the commonly encountered problem areas. These techniques require skills and a depth of fundamental understanding not ordinarily developed in first courses in mechanics.

In the chapters to follow, a variety of the most commonly encountered mechanisms will be examined. They will be classified and procedures for analyzing their behavior described. The procedures presented are not exhaustive. They were selected on the basis of the criterion that the mathematical complexity must not obscure the mechanics principles. Designers are usually not mathematicians, and therefore tend to view things "physically" first, and then mathematically only as need be. It is in that spirit that this text was devised.

2

KINEMATICS

2.1 MACHINE DYNAMICS

Dynamics problems fall into one of two categories, depending on the information given and the results sought. Take, for example, the flight of a spacecraft. The forces that celestial bodies exert on the craft are known. Once the thrusters are built and tested, the forces and moments they can exert are also known. If the craft is launched into drag-free space and the thrusters never used, the equations of motion containing the known celestial forces can be solved for the position of the craft at any time. The resulting path may or may not be the desired one. On the other hand, if the desired trajectory is specified as a function of time, then the equations of motion determine how the thrusters should be used to keep the craft on course.

In the first case, the forces are given and the trajectory is found. In the second, the trajectory and planetary forces are given and the required thruster forces found. Solving the first problem is usually more difficult than the second since it involves twice integrating the equations of motion to find the displacements. When the path is given as a function of time, the accelerations are obtained by differentiation, and then the required forces and moments determined algebraically from the momentum equations. This method is commonly called the *inverse method*.

Most problems in machine dynamics fall into the second category. Mechanical devices are usually designed to produce certain paths, i.e., translating, rotating, or a combination of both. This is done using shafts, cams, gears, links, etc. joined together in an appropriate way. The results of the initial design are usually drawings on paper or on a computer graphic display. The dynamic aspect is added when the device is built and operated. Presumably the machine will execute the desired displacements unless the parts fail. Parts fail because they are overloaded. They may break because they are not strong enough, they may fatigue because of cyclic loading, or sliding surfaces may wear out. Whatever the failure mode, it is ultimately attributable to motion since the worst that can happen to a static structure is that it will decay or corrode away.

The velocities and accelerations of the moving components of a machine designed to execute prescribed displacements can be determined using only the equations of kinematics, without regard for the forces or moments required to produce them. All that is needed to obtain them is a description of the system's possible geometric configurations and adequate information about the velocities and accelerations of some of its members. When the velocities and accelerations of all the members are known, the loads at their connections can be found from the equations of motion. The effects of these loads on each individual member can then be determined using the techniques of strength of materials, elasticity theory, finite elements, or whatever is appropriate to the member being examined.

If the required displacements are known, the analytical processes that follow are, in serial order:

1. Place the mechanism in the configuration of interest.
2. Determine the velocities and the accelerations for these configurations using the kinematic equations and known kinematic inputs.

3. Determine the loads on the elements from the momentum equations using the accelerations.

4. Determine the stresses and strains that result from the loads.

5. Compare these with established failure criteria to ascertain, if possible, the life of the element.

6. Redesign if required and repeat the process.

The material in this chapter and the next are arranged in just this order, beginning with the synthesized mechanism and its kinematics in this chapter, and proceeding through the determination of the loads and the effects that they have on the machine elements in the next. In practice, this is often only the first step in the design process. Necessary adjustments in the original design are made at this point and the process repeated until an acceptable design is obtained. The technique is iterative, partly synthetic, but mostly analytic.

The time is fast approaching when, with the aid of high-speed computers and graphic terminals, the process may be entirely synthetic. By starting with the prescribed motion, mechanisms will be selected, analyzed, and finalized; numerically controlled machines will be programmed, materials selected and ordered, parts made and assembled, all in an automated factory with a minimum of human participation. When that day is reached, the systems that perform these operations will do no more than mimic the work now laboriously performed by humans. Behind each of these marvels will be fundamental concepts of mechanics, some of which are given in the following chapters.

2.2 KINEMATIC EQUATIONS WITH MOVING REFERENCE FRAMES

The kinematic analysis of even simple mechanisms usually involves the simultaneous consideration of several connected elements. The velocities and accelerations of these elements with respect to ground and with respect to each other are often very difficult to visualize, much less analyze. Clearly, what is needed is a systematic way of addressing the problem, preferably one that permits complex problems to be broken down into simpler ones, which can then be investigated piecemeal. Moving reference frames that rotate and have translating origins are well suited to this purpose and for that reason they will be used extensively throughout this text.

There is another very practical reason why moving reference frames must be considered when formulating the kinematics of a system. Very often the performance test data of a machine are measured in terms of noninertial moving-reference-frame coordinates. For example, the speed of an aircraft gas-turbine rotor is indicated by a tachometer attached to the aircraft. When the aircraft is in flight, the turbine speed is referenced to a coordinate system attached to the aircraft. As the aircraft maneuvers, this coordinate system moves with respect to the ground. Newton's laws of motion require that the acceleration be measured with respect to an inertial reference frame, e.g., the ground. Therefore, to be useful, kinematic formulations must be capable of relating noninertial kinematic inputs, such as test data, to an inertial reference frame, usually the earth.

The derivation of the velocity and acceleration expressions for systems using moving reference frames are given in detail in most undergraduate mechanics texts. Although this text concentrates on their applications to mechanisms, an abbreviated derivation is given that highlights the important concepts employed in their development.

It is customary to attach or fix unit vectors $\mathbf{i}$, $\mathbf{j}$, and $\mathbf{k}$ to a moving coordinate system. The derivation of the kinematic equations using auxiliary coordinates requires taking the time derivative of unit vectors $\mathbf{i}$, $\mathbf{j}$, and $\mathbf{k}$ since their orientation changes with time. The formulation of this derivative can be illustrated by examining the two-dimensional case shown in Figure 2.1, where vectors $\mathbf{i}$ and $\mathbf{j}$ are constrained to move in the plane of fixed coordinates X and Y.

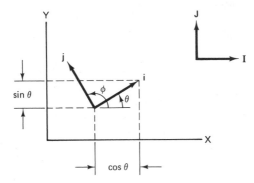

Figure 2.1   Moving Unit Vectors

In terms of the fixed-coordinate-base vectors $\mathbf{I}$, $\mathbf{J}$, and $\mathbf{K}$,

$$\mathbf{i} = (\cos \theta)\mathbf{I} + (\sin \theta)\mathbf{J}$$

The time derivative of this unit vector is

$$d\mathbf{i}/dt = -\dot{\theta}(\sin \theta)\mathbf{I} + \dot{\theta}(\cos \theta)\mathbf{J}$$

This expression can be obtained by the direct expansion of the cross product $\dot{\theta}\mathbf{K} \times \mathbf{i}$ so that the last derivative can also be written as

$$d\mathbf{i}/dt = \dot{\theta}\mathbf{K} \times \mathbf{i}$$

In words, this equation states that the time derivative of a unit vector with respect to a fixed coordinate system is the cross product of its angular velocity (measured with respect to that system) and itself.

In like manner,

$$d\mathbf{j}/dt = \dot{\phi}\mathbf{K} \times \mathbf{j}$$

If the two unit vectors maintain a fixed angle between them, e.g., a right angle, then they have a common angular velocity $\dot{\phi}$ or $\dot{\theta}$.

Figure 2.2 shows a three-dimensional fixed or inertial reference frame, XYZ, and a moving reference frame, xyz.

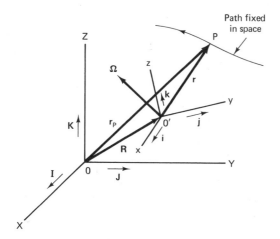

Figure 2.2   Moving Reference Frame

The coordinate axes of both systems form rigid triads that are always orthogonal to each other and follow the right-hand rule for their ordering. Since the moving coordinates xyz behave as a rigid body, they have a common angular velocity $\boldsymbol{\Omega}$. The derivatives of the unit vectors $\mathbf{i}$, $\mathbf{j}$, and $\mathbf{k}$ fixed in this system are given by

$$d\mathbf{i}/dt = \boldsymbol{\Omega} \times \mathbf{i} \qquad d\mathbf{j}/dt = \boldsymbol{\Omega} \times \mathbf{j} \qquad d\mathbf{k}/dt = \boldsymbol{\Omega} \times \mathbf{k} \qquad (2.1)$$

The symbol $\boldsymbol{\Omega}$ will be reserved for the angular velocity of the coordinates xyz, and $\boldsymbol{\omega}$ will be used for the angular velocity of rigid bodies such as links, cams, and shafts. It should be noted that $\boldsymbol{\Omega}$ can be written in terms of $\mathbf{i}$, $\mathbf{j}$, and $\mathbf{k}$ or $\mathbf{I}$, $\mathbf{J}$, and $\mathbf{K}$, whichever way makes the analysis of the problem easier. It will often be convenient to fix xyz to one of the members of the system, but that is not required.

The vector $\mathbf{R}$ shown in Figure 2.2 locates the origin of the moving coordinate system. The vector $\mathbf{r}$ locates the point of interest, P, relative to the moving coordinate system. The components of this vector are its projections on to the xyz coordinates. Point P is said to be fixed in the moving coordinate system if the projections of $\mathbf{r}$ in the moving coordinate system do not change with time. In that case, there would be no relative motion of point P with respect to the xyz coordinate system. Whether or not there is any relative motion depends on the motion assigned to the moving coordinate system. For example, the point might be fixed in the XYZ system and yet move relative to xyz because the xyz system is translating, or rotating, or both.

When base vectors are used, it is convenient to express the location of point P relative to xyz in terms of its coordinates (also called xyz) and the unit base vectors $\mathbf{i}$, $\mathbf{j}$, and $\mathbf{k}$ fixed in the xyz system. Thus,

$$\mathbf{r} = x\mathbf{i} + y\mathbf{j} + z\mathbf{k} \qquad (2.2)$$

The position vector of point P can now be written, according to Figure 2.2, in the form

$$\mathbf{r}_P = \mathbf{R} + \mathbf{r}$$

Differentiating this vector expression with respect to time yields

$$\mathbf{V} = d\mathbf{r}_P/dt = \dot{\mathbf{R}} + \dot{\mathbf{r}}$$

where the dot notation means total differentiation with respect to time, i.e., $(\dot{}) = d/dt$. The vector $\dot{\mathbf{R}}$ is the velocity of the origin of the auxiliary coordinate system. The second term on the right side is given by

$$\dot{\mathbf{r}} = x(d\mathbf{i}/dt) + y(d\mathbf{j}/dt) + z(d\mathbf{k}/dt) + \dot{x}\mathbf{i} + \dot{y}\mathbf{j} + \dot{z}\mathbf{k}$$

The first three terms in this expression can be rewritten

$$x(d\mathbf{i}/dt) + y(d\mathbf{j}/dt) + z(d\mathbf{k}/dt) = \mathbf{\Omega} \times \mathbf{r}$$

by the direct substitution of the unit-vector derivative formulas. This term represents the contribution to the velocity made by $\mathbf{r}$ if it were fixed in xyz and rotating with it at $\mathbf{\Omega}$.

The second three terms are the relative velocity of P with respect to the moving coordinates. The symbol for this contribution is $\mathbf{v}_{xyz}$ so that

$$\mathbf{v}_{xyz} = \dot{x}\mathbf{i} + \dot{y}\mathbf{j} + \dot{z}\mathbf{k} \tag{2.3}$$

With these, the velocity equation becomes

$$\mathbf{V} = \dot{\mathbf{R}} + \mathbf{\Omega} \times \mathbf{r} + \mathbf{v}_{xyz} \tag{2.4}$$

As it stands, this is a very sterile equation. The chances of solving even the simplest kinematic problem not knowing the precise meaning of each term in this equation are very slim indeed. Word descriptions of each term are summarized in the list that follows and should be thoroughly learned by the student before attempting any problem.

1. $\mathbf{V}$ is the velocity of the point of interest with respect to the XYZ coordinates.
2. $\dot{\mathbf{R}}$ is the translational velocity of the origin of the xyz system chosen by the user.
3. $\mathbf{\Omega}$ is the angular velocity of the xyz system with respect to the XYZ system chosen by the user.
4. $\mathbf{r}$ locates the point of interest in the xyz system.
5. $\mathbf{v}_{xyz}$ is the velocity of the point of interest relative to the xyz coordinate system. The components of this vector cannot be determined until $\dot{\mathbf{R}}$ and $\mathbf{\Omega}$ are chosen. It will be necessary to include this term whenever the projections of $\mathbf{r}$ onto the xyz coordinates change with time.

Students will find that many misapplications of the velocity equation can be avoided if they actually say the words describing the individual terms as they apply them.

The acceleration of P is obtained by a continuation of the differentiation process, i.e.,

$$\mathbf{A} = \ddot{\mathbf{R}} + \dot{\mathbf{\Omega}} \times \mathbf{r} + \mathbf{\Omega} \times \dot{\mathbf{r}} + \dot{\mathbf{v}}_{xyz}$$

The third term on the right contains $\dot{\mathbf{r}}$, which was examined earlier. It can be written as

$$\boldsymbol{\Omega} \times \dot{\mathbf{r}} = \boldsymbol{\Omega} \times (\boldsymbol{\Omega} \times \mathbf{r} + \mathbf{v}_{xyz}) = \boldsymbol{\Omega} \times (\boldsymbol{\Omega} \times \mathbf{r}) + \boldsymbol{\Omega} \times \mathbf{v}_{xyz}$$

The fourth term on the right becomes

$$\dot{\mathbf{v}}_{xyz} = \ddot{x}\mathbf{i} + \ddot{y}\mathbf{j} + \ddot{z}\mathbf{k} + \dot{x}\,d\mathbf{i}/dt + \dot{y}\,d\mathbf{j}/dt + \dot{z}\,d\mathbf{k}/dt$$

The first three terms of this expression represent the relative acceleration, i.e.,

$$\mathbf{a}_{xyz} = \ddot{x}\mathbf{i} + \ddot{y}\mathbf{j} + \ddot{z}\mathbf{k} \tag{2.5}$$

The last three terms reduce to

$$\dot{x}(\boldsymbol{\Omega} \times \mathbf{i}) + \dot{y}(\boldsymbol{\Omega} \times \mathbf{j}) + \dot{z}(\boldsymbol{\Omega} \times \mathbf{k}) = \boldsymbol{\Omega} \times \mathbf{v}_{xyz}$$

With these substitutions, the acceleration becomes

$$\mathbf{A} = \ddot{\mathbf{R}} + \mathbf{a}_{xyz} + 2\boldsymbol{\Omega} \times \mathbf{v}_{xyz} + \dot{\boldsymbol{\Omega}} \times \mathbf{r} + \boldsymbol{\Omega} \times (\boldsymbol{\Omega} \times \mathbf{r}) \tag{2.6}$$

The quantities $\boldsymbol{\Omega}$, $\mathbf{v}_{xyz}$, and $\mathbf{r}$ have the same word descriptions as before. The new terms are

1. $\ddot{\mathbf{R}}$, the acceleration of the origin of the moving coordinate system.
2. $\mathbf{a}_{xyz}$, the acceleration of the point of interest relative to the xyz coordinate system. This term cannot be determined until $\dot{\mathbf{R}}$, $\ddot{\mathbf{R}}$, $\boldsymbol{\Omega}$, and $\dot{\boldsymbol{\Omega}}$ are chosen. In terms of the moving base vectors $\mathbf{i}$, $\mathbf{j}$, and $\mathbf{k}$, the relative acceleration is formulated as $\mathbf{a}_{xyz} = \ddot{x}\mathbf{i} + \ddot{y}\mathbf{j} + \ddot{z}\mathbf{k}$.
3. $\dot{\boldsymbol{\Omega}}$ is the angular acceleration of the xyz coordinate system with respect to XYZ.

The term $2\boldsymbol{\Omega} \times \mathbf{v}_{xyz}$ is called the Coriolis acceleration in honor of the man who discovered it. Its appearance in the equation is a direct result of having used a noninertial rotating reference system xyz. The clockwise circulation around high-pressure weather systems and the counterclockwise circulation around low-pressure systems is caused by this term, a clear demonstration that earthbound coordinate systems cannot be considered as inertial reference systems for very large-scale systems. For the size systems that we will deal with, XYZ coordinates attached to ground can be considered to be "inertial."

The terms $\boldsymbol{\Omega} \times (\boldsymbol{\Omega} \times \mathbf{r})$ and $\dot{\boldsymbol{\Omega}} \times \mathbf{r}$ are related to the familiar centripetal and tangential accelerations. The first of these is always directed inward along $\mathbf{r}$, hence the name centripetal. The second is always normal to $\mathbf{r}$ and also to $\dot{\boldsymbol{\Omega}}$. Both terms depend on the motion assigned to the xyz system and cannot be determined until that choice is made. Bear in mind that $\boldsymbol{\Omega}$ and $\dot{\boldsymbol{\Omega}}$ used in the velocity and acceleration equations are "absolute" quantities in the sense that they are measured with reference to the fixed XYZ coordinates.

The velocity and acceleration equations as we use them will describe these two kinematic quantities only at an instant of time. They will not predict the future or the past. The velocities and accelerations for various possible configurations of the system are found by the reapplication of these equations to each new configuration.

One disadvantage of an instantaneous analysis is that it does not reveal the time at which the velocity and acceleration occur. Fortunately, this does not diminish the usefulness of the results. Kinematic analyses are usually performed as a prelude to the evaluation of the inertial loads on the system. Members of the system must survive these loads regardless of when they occur.

2.3 RADIUS OF CURVATURE AND PATH COORDINATES

The application of the kinematic equations invariably requires the selection of one or more moving coordinate systems. Deciding where to place these systems and how to assign their motions is often the most difficult part of the analysis. The equations themselves do not indicate how to make these decisions, and there are no rules built into the derivation of the equations that would dictate how they are to be used. The user must decide, depending on the application, what coordinate motions are best suited to the solution of that particular problem.

Fortunately, there is a practical basis for making these selections. The clue is contained in the relative velocity and relative acceleration terms, i.e., $\mathbf{v}_{xyz}$ and $\mathbf{a}_{xyz}$. These terms are determined by the choice of the moving coordinates system, i.e., by the selection of $\dot{\mathbf{R}}$, $\ddot{\mathbf{R}}$, $\mathbf{\Omega}$, and $\dot{\mathbf{\Omega}}$. Different choices yield different $\mathbf{v}_{xyz}$ and $\mathbf{a}_{xyz}$. Choosing an xyz system in which it is difficult or impossible to forumlate $\mathbf{v}_{xyz}$ and $\mathbf{a}_{xyz}$ makes little sense. Only those xyz coordinate systems that permit one to visualize and thus formulate the relative motions warrant consideration.

There will always be at least one choice of moving coordinates that will permit the relative motions to be viewed as a curved (or straight-line) displacement of a point fixed on one member relative to an adjoining member. The relative velocities and accelerations of these motions are most conveniently described using the path coordinates. To illustrate the use of these coordinates consider the kinematic problem illustrated in Figure 2.3.

The plate in this figure is translating in the plane of the page and rotating at the same time. Moving coordinate axes have been fixed to the plate at O and have been assigned $\mathbf{\Omega} = \boldsymbol{\omega}$ and $\dot{\mathbf{\Omega}} = \dot{\boldsymbol{\omega}}$. As a result, the slot in the plate is also fixed in the xy coordinate system. The pin, which is attached to an adjacent member, slides in the slot. As far as the relative motion is concerned, an observer fixed in the xy system sees the pin moving along a curved path that remains fixed relative to his or her xy system. He or she is not concerned with the motion of the plate with respect to the XY system as these motions are taken into account by the other terms in the kinematic equations. By attaching the moving xy coordinate system to the plate, the simplest relative motion possible has been created, i.e., along a path that is fixed in this system.

As the name implies, path coordinates are well suited to the formulation of the velocity and acceleration of a point moving along a predetermined path that is fixed

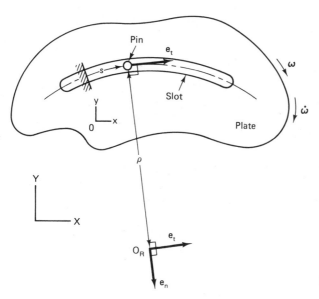

Figure 2.3   Path Coordinates

in that coordinate system. When path coordinates are used, a unit-vector tangent $\mathbf{e}_t$ to the path is attached to the moving point and the velocity written as

$$\mathbf{v}_{xyz} = \dot{s}\mathbf{e}_t \tag{2.7}$$

The scalar coefficient in this formulation, $\dot{s}$, is the speed of the pin measured along the curved path fixed to the plate. The unit-vector tangent $\mathbf{e}_t$ turns with the path as the pin moves along the path, so that the differentiation of the velocity equation with respect to time requires differentiating $\mathbf{e}_t$. This vector has a derivative, not because its length changes, but because its direction changes. The rate at which it changes depends on how fast the point moves along the path and the curvature of the path. For those reasons, the equation for the acceleration contains two terms, one that accounts for the acceleration along the path, $\ddot{s}\mathbf{e}_t$, and one that accounts for the rate at which $\mathbf{e}_t$ turns, $(\dot{s}^2/\rho)\mathbf{e}_n$. The result is

$$\mathbf{a}_{xyz} = \ddot{s}\mathbf{e}_t + (\dot{s}^2/\rho)\mathbf{e}_n \tag{2.8}$$

where ρ is the "radius of curvature" shown in Figure 2.3.

The second term on the right-hand side of this equation, which results from differentiating $\mathbf{e}_t$, has yielded a new unit vector, $\mathbf{e}_n$. This vector is normal to the path on the concave side of the curve, pointing in the direction of change of $\mathbf{e}_t$. The scalar $\dot{s}/\rho$ measures the rate of this change, i.e., the instantaneous angular velocity of the unit vector $\mathbf{e}_t$ as it progresses along the path. The orthogonal vectors $\mathbf{e}_t$ and $\mathbf{e}_n$ define the path coordinate directions, which are always tangent and normal to the path.

The radius of curvature ρ, a scalar, is a property of the curve, independent of the motion. This is evident from the following formula that is used to compute ρ in two-dimensional space when the equation for the curve is known and is twice differentiable.

$$\rho = [1 + (dy/dx)^2]^{3/2} / |d^2y/dx^2| \qquad (2.9)$$

In practice, the equation for the curve may not be available. In that case, the radius of curvature may be shown on the drawing of the part. If not, a reasonably good estimate can be found by constructing several normals to the curve at and nearby the point under consideration. The intersection of these locates O_R in Figure 2.3.

Figure 2.4 illustrates a novel method for constructing these normals. A reflecting surface, such as a mirror, is placed at the point and rotated about an axis normal to the plane of the curve until the reflected image is a smooth continuation of the drawn curve. In this position, the edge of the mirror is normal to the curve.

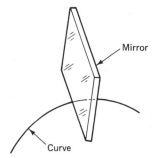

Figure 2.4 Method for Finding Normal to any Curve with a Mirror

No matter which method is used, the chance that more than two construction lines will cross at one common point is highly unlikely, so that some judgment will usually be required when using graphical methods.

The methods described are not only practical, but also serve to illustrate an important property of the radius of curvature. Obviously, the radius of curvature cannot be constructed from a single normal to the curve emanating from the point of interest. At least two normals are needed to obtain an intersection and they must emanate from different but close by points on the curve. From a strictly mathematical viewpoint, one normal is not sufficient because the curvature of a curve at a point depends on the slopes of the curve adjacent to that point. The second derivative in the formula for ρ reflects this "nearby" feature of the concept. It measures the change in the slope of the curve, i.e., the difference in the slopes at small distances either side of the point in the limit as these distances are reduced toward zero. There are curves for which a single-valued result could not be obtained this way. For example, a curve with a cusp does not even have a continuous first derivative. Pathological curves such as this rarely if ever occur in practice.

A somewhat more simplistic notion of the concept can be obtained by viewing the radius of curvature as the result of "curve fitting" a small portion of a circle to a small portion of the real path. Since there are many circles that can be drawn tangent to a given path at some point (Figure 2.5), tangency is not sufficient to define

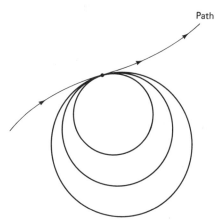

Figure 2.5 Circles Tangent to a Path

ρ. Equation (2.9) shows that of these circles, the "circle of curvature" is the one that has both its first and second derivatives equal to the first and second derivatives of the curve at the point. This particular circle has a higher degree of contact with the path than any other circle. It is often referred to as an osculating circle and its radius is the radius of curvature.

From a practical standpoint, this means that a small portion of the path in the neighborhood of the point may be considered to be an arc of a perfect circle with a radius ρ. If the path itself is not a perfect circle, the size and direction of the radius of curvature will be different for every point along the path. The important thing to remember is that the path-coordinate formulation of the instantaneous acceleration considers ρ to be a constant for a very small portion of the path nearby the point. As far as the velocity and acceleration equations are concerned, ρ is a constant at any instant. If it were not, the time derivative of ρ would appear in the path equations for velocity and acceleration.

The instantaneous constancy of the radius of curvature is not only a concept fundamental to the formulation of the acceleration vector, it also permits the construction of "equivalent mechanisms" that greatly simplify the analysis of contacting members. The methods for identifying and using equivalent mechanisms will be developed later in this chapter. Suffice to say that the radius of curvature is an exceedingly powerful geometric construct that is used frequently in the kinematic analysis of mechanisms.

2.4 VELOCITIES OF FOUR-BAR LINKAGES

Figure 2.6 shows a four-bar mechanism consisting of a bell-crank (2), a connecting rod (3), and a rotating disk (4). The connecting rod is pinned to the bell-crank and disk at points A and B. The bell-crank is pinned to ground (1) at O_2 and rotates at a constant angular velocity of $\omega_2 = 1.0$ rad/s in the clockwise direction. The disk (4)

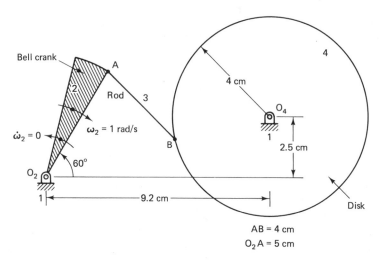

AB = 4 cm

O_2A = 5 cm

Figure 2.6   Bell-Crank Mechanism

is also pinned to ground (1) at O_4. The ground (1) acts as the fourth member and is stationary.

The kinematic determination of the velocities and accelerations will be obtained using graphical methods. We will start with member (2), the bell-crank, since that is the member we know most about. Figure 2.7 shows the member separated from the mechanism. Since the shape of this member plays no role in its kinematics, it has been replaced by the line O_2A.

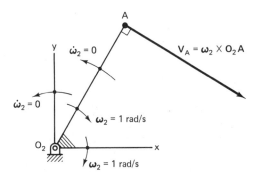

Figure 2.7   Bell-Crank Mechanism with Attached Coordinate

A rotating coordinate system has been attached to O_2 and assigned the angular velocity and angular acceleration of member (2). As a result, projections of O_2A onto the x and y coordinates do not change with time because vector $\mathbf{O_2A}$ has been fixed in the xy coordinates by their assigned motion. For this moving system, $\mathbf{R} = \mathbf{v}_{xy} = \mathbf{0}$, $\mathbf{\Omega} = \mathbf{\omega}_2$, and $\mathbf{r} = O_2A$. The velocity of A is, therefore,

$$\mathbf{V}_A = \mathbf{\omega}_2 \times \mathbf{O_2A}$$

This vector, shown in Figure 2.7, is normal to vector $\mathbf{O_2A}$. Because $\mathbf{\omega}_2$ and $\mathbf{O_2A}$

are perpendicular, its length is equal to $\omega_2(O_2A) = 5$ cm/s and it is directed according to the right-hand rule.

Having found the velocity of point A, we will now try to find the velocity of B using another moving coordinate system, this time attached to A. This new set of coordinates is not related in any way to the previous ones. It will be assigned the angular velocity and angular acceleration of member (3), and its origin will have the velocity and acceleration of point A.

Figure 2.8 shows member (3) with the new xy coordinates attached. For this coordinate system, $\dot{\mathbf{R}} = \mathbf{V}_A$, $\mathbf{\Omega} = \boldsymbol{\omega}_3$, $\dot{\mathbf{\Omega}} = \dot{\boldsymbol{\omega}}_3$, and $\mathbf{r} = \mathbf{AB}$. The velocity of point B is given by

$$\mathbf{V}_B = \mathbf{V}_A + \boldsymbol{\omega}_3 \times \mathbf{AB}$$

Vectors $\mathbf{V}_A$ and $\mathbf{AB}$ are known. The only thing that can be said about $\boldsymbol{\omega}_3$ is that it is normal to the plane of the mechanism. That means that $\boldsymbol{\omega}_3 \times \mathbf{AB}$ is normal to $\mathbf{AB}$ in the plane of the page, directed away from that line on one side or the other.

We are now ready to begin the construction of the velocity polygon. Figure 2.9 shows a construction line drawn parallel to link O_2A. The velocity $\mathbf{V}_A$ is shown normal to this line drawn to some convenient scale. The origin of the velocity polygon has been labeled O_v to reference the zero-velocity point in the velocity plane.

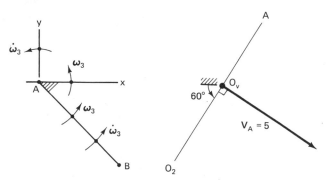

Figure 2.8   Rod with Attached Coordinates

Figure 2.9   Construction of Velocity of A

According to the velocity equation, $\boldsymbol{\omega}_3 \times \mathbf{AB}$ is to be added vectorially to $\mathbf{V}_A$ to get $\mathbf{V}_B$. At this stage of the analysis, the most that can be said about this vector is that it is normal to AB. Figure 2.10 shows a construction line parallel to AB drawn through the end of vector $\mathbf{V}_A$. Another construction line is shown normal to AB. The vector $\boldsymbol{\omega}_3 \times \mathbf{AB}$ lies along this line, with its tail at the end of $\mathbf{V}_A$. The actual construction of this vector cannot be finished now since we do not know which side of the line to draw it on or how long to draw it.

The velocity of B can be formulated by placing an xy coordinate system at O_4 and letting it rotate with member (4), as shown in Figure 2.11. Then the velocity of B can be expressed as

$$\mathbf{V}_B = \boldsymbol{\omega}_4 \times \mathbf{O}_4\mathbf{B}$$

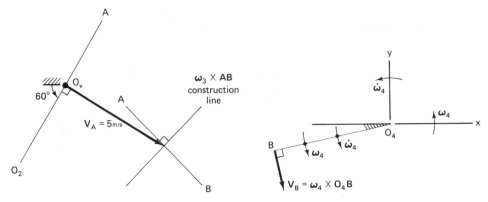

Figure 2.10   Construction Line for $\omega_3 \times$ **AB**   **Figure 2.11**   Sketch of **V**$_B$ Construction

This is the same formulation used to find **V**$_A$ except that ω_4 is an unknown. We do know that **V**$_B$ is normal to **O**$_4$**B**. Although the angular velocity and acceleration have been shown in the positive sense in Figure 2.11, this in no way anticipates the correct sense of these vectors. Those will be found from the graphical solution.

Figure 2.10 has been redrawn in Figure 2.12 with a construction line parallel to O_4B shown passing through O_v. Another construction line has been drawn through O_v normal to O_4B. The vector $\omega_4 \times$ **O**$_4$**B** lies along this line since it is perpendicular to O_4B.

The intersection of the $\omega_4 \times$ **O**$_4$**B** construction line with the $\omega_3 \times$ **AB** construction line determines the size and sense of both $\omega_4 \times$ **O**$_4$**B** and $\omega_3 \times$ **AB**. The

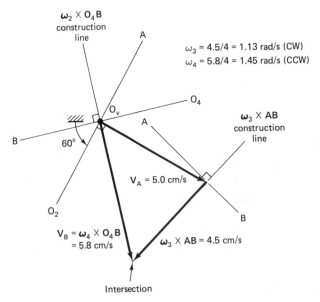

$\omega_3 = 4.5/4 = 1.13$ rad/s (CW)
$\omega_4 = 5.8/4 = 1.45$ rad/s (CCW)

Figure 2.12   Completed Velocity Diagram

intersection of these lines shown in Figure 2.12 is equivalent to the simultaneous solution of the two scalar equations that would result from the unit-vector formulation
of the problem. The angular velocities can now be determined from the scaled values of the intersecting vectors.

The sense of the angular velocities can be determined by visualizing the motion an observer at the origin of each coordinate system would see. An observer at O_4
would see a counterclockwise rotation (CCW) of B attached to member (4). An observer at A would see rod AB rotating clockwise, therefore, ω_3 is clockwise (CW).
These results are consistent with what intuition might suggest. Unfortunately, intuition is often an unreliable guide to the sense of angular acceleration vectors.

The sense of the angular velocity can also be found by guessing the direction
of ω, forming $\omega \times r$ by the right-hand rule, and comparing the result with the velocity diagram. If the direction of $\omega \times r$ agrees with the diagram, the direction of
ω was guessed correctly.

The acceleration analysis will follow the same strategy as the velocity analysis.
The same three moving xy coordinates will be used so that all the velocities obtained
from the previous analysis still apply. Each xy system will be assigned the angular
velocity and angular acceleration of the member to which it is attached. Each origin
will accelerate with its point of attachment.

The acceleration of point A is given by

$$\mathbf{A}_A = \ddot{\mathbf{R}} + \mathbf{a}_{xyz} + 2\boldsymbol{\Omega} \times \mathbf{v}_{xyz} + \boldsymbol{\Omega} \times (\boldsymbol{\Omega} \times \mathbf{r}) + \dot{\boldsymbol{\Omega}} \times \mathbf{r}$$

For an xy coordinate system with its origin at O_2 rotating with member (2),

$$\ddot{\mathbf{R}} = \mathbf{0} \qquad \mathbf{v}_{xyz} = \mathbf{0}$$

$$\mathbf{a}_{xyz} = \mathbf{0} \qquad \boldsymbol{\Omega} = \boldsymbol{\omega}_2$$

$$\mathbf{r} = O_2\mathbf{A} \qquad \dot{\boldsymbol{\Omega}} = \dot{\boldsymbol{\omega}}_2 = \mathbf{0}$$

Therefore,

$$\mathbf{A}_A = \boldsymbol{\omega}_2 \times (\boldsymbol{\omega}_2 \times O_2\mathbf{A})$$

Application of the right-hand rule for cross products will show that this is a centripetal acceleration directed from A toward O_2.

Generally, point A would be expected to have both centripetal and tangential
accelerations since A moves on a circular path about O_2. The centripetal, or normal,
component of acceleration would be

$$\mathbf{A}_A^n = \boldsymbol{\omega}_2 \times (\boldsymbol{\omega}_2 \times O_2\mathbf{A})$$

and the tangential component would be

$$\mathbf{A}_A^t = \dot{\boldsymbol{\omega}}_2 \times O_2\mathbf{A}$$

The total acceleration would be the vector sum of these, i.e.,

$$\mathbf{A}_A = \mathbf{A}_A^n + \mathbf{A}_A^t$$

Because $\dot{\omega}_2 = 0$ in this case, the total acceleration of A is just the normal (centripetal) component.

For an xy coordinate system with an origin at A rotating with member (3),

$$\ddot{\mathbf{R}} = \mathbf{A}_A \qquad \mathbf{v}_{xyz} = \mathbf{0}$$

$$\mathbf{a}_{xyz} = \mathbf{0} \qquad \mathbf{\Omega} = \boldsymbol{\omega}_3$$

$$\mathbf{r} = \mathbf{AB} \qquad \dot{\mathbf{\Omega}} = \dot{\boldsymbol{\omega}}_3$$

Then,

$$\mathbf{A}_B = \mathbf{A}_A + \boldsymbol{\omega}_3 \times (\boldsymbol{\omega}_3 \times \mathbf{AB}) + \dot{\boldsymbol{\omega}}_3 \times \mathbf{AB}$$

Vector $\mathbf{A}_A$ is already known. The size of centripetal term $\boldsymbol{\omega}_3 \times (\boldsymbol{\omega}_3 \times \mathbf{AB})$ can be determined from the velocity diagram. Its direction is along AB from B toward A. The tangential term $\dot{\boldsymbol{\omega}}_3 \times \mathbf{AB}$ is perpendicular to AB, but its sense and size are unknown. Figure 2.13 shows the acceleration diagram with the known vectors drawn on it. The construction line normal to line AB indicates the line of action of the unknown vector $\dot{\boldsymbol{\omega}}_3 \times \mathbf{AB}$. Point O_a is the origin of the acceleration polygon in the acceleration plane.

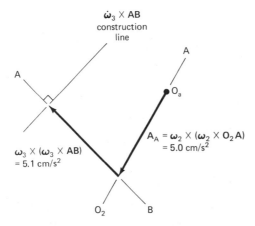

Figure 2.13   Centripetal Accelerations

The acceleration of point B, formulated using an xy coordinate system fixed at O_4 and rotating with member (4), has the following inputs:

$$\ddot{\mathbf{R}} = \mathbf{0} \qquad \mathbf{v}_{xyz} = \mathbf{0}$$

$$\mathbf{a}_{xyz} = \mathbf{0} \qquad \mathbf{\Omega} = \boldsymbol{\omega}_4$$

$$\mathbf{r} = \mathbf{O}_4\mathbf{B} \qquad \dot{\mathbf{\Omega}} = \dot{\boldsymbol{\omega}}_4$$

The acceleration of B is given by

$$\mathbf{A}_B = \boldsymbol{\omega}_4 \times (\boldsymbol{\omega}_4 \times \mathbf{O}_4\mathbf{B}) + \dot{\boldsymbol{\omega}}_4 \times \mathbf{O}_4\mathbf{B}$$

Point B follows a circular path about O_4, but not necessarily with a constant angular velocity. The centripetal acceleration $\omega_4 \times (\omega_4 \times O_4 B)$ can be evaluated from the results of the velocity diagram. It is directed toward O_4 from B. The tangential component $\dot{\omega}_4 \times O_4 B$ is normal to $O_4 B$. Figure 2.14 shows the acceleration diagram of Figure 2.13 with $\omega_4 \times (\omega_4 \times O_4 B)$ drawn from the origin O_a parallel to $O_4 B$. The $\dot{\omega}_4 \times O_4 B$ construction line drawn normal to $O_4 B$ is also shown.

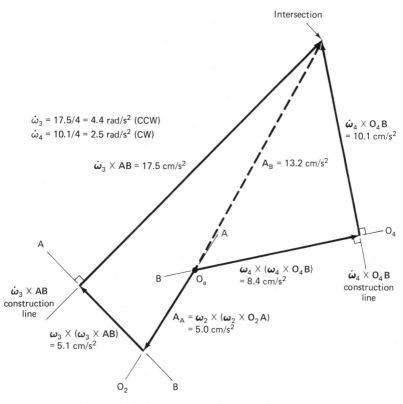

Figure 2.14   Completed Acceleration Diagram

The intersection of the construction lines for $\dot{\omega}_4 \times O_4 B$ and $\dot{\omega}_3 \times AB$ in this figure determines the sense and magnitude of these tangential components. The angular accelerations $\dot{\omega}_3$ and $\dot{\omega}_4$ can be calculated from these results. Observed from O_4, $\dot{\omega}_4$ is clockwise. Observed from A, $\dot{\omega}_3$ is counterclockwise. The acceleration of B is shown by a dashed line emanating from O_a.

The kinematic analysis of the four-bar mechanism is now complete. Velocities and accelerations for other possible configurations can be found by repeating the process.

2.5 SLIDING CONTACT: STRAIGHT-LINE RELATIVE MOTION

In the previous example, all the members were joined by pins. Figure 2.15 shows an escapement, or indexing mechanism, that is commonly called a Geneva wheel. A pin at point P_2 on the perimeter of disk (2) periodically engages with the slots on the star (3), causing it to index to one of its four stationary positions. A raised, circular ridge on the face of the star mates with the circular edge of the wheel to maintain its location when it is not being indexed. During indexing, the pin slides into and out of the slots as it rotates the star. Unlike the previous mechanism, there is a relative motion of the two members at their point of contact. As a consequence, the terms $\mathbf{v}_{xyz}$ and $\mathbf{a}_{xyz}$ will appear in the velocity and acceleration equations.

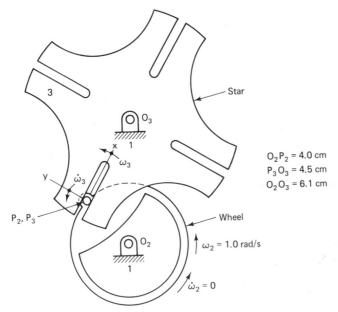

$$O_2P_2 = 4.0 \text{ cm}$$
$$P_3O_3 = 4.5 \text{ cm}$$
$$O_2O_3 = 6.1 \text{ cm}$$

Figure 2.15   Geneva Escapement

To facilitate the analysis of the relative motion, it will be convenient to give names to two points that are coincident in Figure 2.15. Point P_2 is at the center of the pin fixed to member (2). Point P_3 is fixed on the center line of the slot and coincident with P_2 at this instant. Points P_2 and P_3 rotate on their respective circles about O_2 and O_3 and are coincident only at this instant.

The relative motion of the pin through the slot is measured by the relative motion of P_2 along the slot center line relative to P_3 fixed on the slot center line. If an xy coordinate system were attached to the slot at P_3 and assigned the angular velocity and acceleration of the slot, the path of the pin relative to this coordinate system would be a straight line as long as the pin was engaged with the slot. The quantities $\dot{x}$ and $\dot{y}$ would be the components of the velocity of P_2 relative to this system, and $\ddot{x}$ and $\ddot{y}$ the components of its relative acceleration. The relative acceleration would be parallel to the slot since the radius of curvature of the path is infinite.

This moving coordinate system is not the only one that might have been chosen. It was selected because the relative velocity and acceleration of the pin along the slot are easy to visualize and formulate when viewed from this system. The student is encouraged to try other systems and verify that the system chosen is the most convenient.

The relative motion was considered first because it is the hardest part of the problem. Once the method of describing the relative motions is established through the selection of moving coordinate systems, the strategy for solving the problem is set. In this case, we will need to formulate the velocity and acceleration of the origin of the moving coordinate system attached to P_3. From Figure 2.15, these are

$$\mathbf{V}_{P3} = \boldsymbol{\omega}_3 \times \mathbf{O}_3\mathbf{P}_3$$

and

$$\mathbf{A}_{P3} = \boldsymbol{\omega}_3 \times (\boldsymbol{\omega}_3 \times \mathbf{O}_3\mathbf{P}_3) + \dot{\boldsymbol{\omega}}_3 \times (\mathbf{O}_3\mathbf{P}_3)$$

Although not stated, these equations imply that an xy system rotating with member (3) has been established with its origin at O_3.

The velocity and acceleration of P_2 are formulated in a similar fashion, i.e.,

$$\mathbf{V}_{P2} = \boldsymbol{\omega}_2 \times \mathbf{O}_2\mathbf{P}_2$$

and since $\dot{\boldsymbol{\omega}}_2 = \mathbf{O}$,

$$\mathbf{A}_{P2} = \boldsymbol{\omega}_2 \times (\boldsymbol{\omega}_2 \times \mathbf{O}_2\mathbf{P}_2)$$

The connection between the two members is made by the velocity equation

$$\mathbf{V}_{P2} = \mathbf{V}_{P3} + \mathbf{v}_{xy}$$

and the acceleration equation

$$\mathbf{A}_{P2} = \mathbf{A}_{P3} + \mathbf{a}_{xy} + 2\boldsymbol{\omega}_3 \times \mathbf{v}_{xy}$$

The three terms in these equations that contain $\mathbf{r}$ are absent because points P_2 and P_3 are coincident at this instant, i.e., $\mathbf{r} = \mathbf{O}$. Because the auxiliary coordinate equation is fixed to the slot at P_3, the path of P_2 in this system is a straight line. The Coriolis term contains the angular velocity of this system, $\boldsymbol{\omega}_3$.

In this problem and in future problems, it will be convenient to adopt the notation $\mathbf{v}_{xy} = \mathbf{v}_{P2P3}$ and $\mathbf{a}_{xyz} = \mathbf{a}_{P2P3}$, where "$P_2P_3$" is read as "$P_2$ relative to P_3." With this notation, the connecting equations are

$$\mathbf{V}_{P2} = \mathbf{V}_{P3} + \mathbf{v}_{P2P3}$$

$$\mathbf{A}_{P2} = \mathbf{A}_{P3} + \mathbf{a}_{P2P3} + 2\boldsymbol{\omega}_3 \times \mathbf{v}_{P2P3}$$

The graphical analyses of the mechanism can now proceed. Figure 2.16 shows the velocity diagram of the system.

Velocity $\mathbf{V}_{P2}$ is drawn first, beginning at the origin of the velocity diagram, O_v, and normal to the construction line O_2P_2. Velocity $\mathbf{V}_{P3}$ is perpendicular to O_3P_3, but

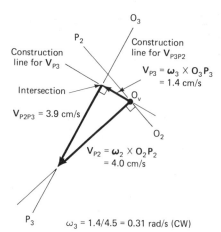

O_3

P_2

Construction
line for V_{P3P2}

Construction
line for V_{P3}

$V_{P3} = \omega_3 \times O_3 P_3$
    $= 1.4$ cm/s

Intersection

O_v

$V_{P2P3} = 3.9$ cm/s

O_2

$V_{P2} = \omega_2 \times O_2 P_2$
    $= 4.0$ cm/s

P_3 $\omega_3 = 1.4/4.5 = 0.31$ rad/s (CW)

Figure 2.16   Velocity Diagram

its sense and magnitude are unknown. A construction line through O_v normal to $O_3 P_3$ is drawn to indicate the line of action of V_{P3}.

The relative velocity v_{P2P3} is known to be parallel to the slot, i.e., parallel to $O_3 P_3$ and perpendicular to V_{P3}. When it is added to the V_{P2} vector, the result is supposed to be V_{P3}. With this in mind, the diagram is closed by constructing a line parallel to $O_3 P_3$ through the tip of the V_{P2} vector. The intersection of this line with the construction line for V_{P3} shown in Figure 2.16 is the simultaneous solution for the two unknown velocities, V_{P3} and v_{P2P3}.

The acceleration diagram follows in the same way. Figure 2.17 shows the necessary constructions.

By starting at the origin O_a, the acceleration of P_2 is directed from P_2 toward O_2 along the construction line $P_2 O_2$. The normal acceleration of P_3, A_{P3}^n, is directed from P_3 toward O_3 along $P_3 O_3$.

The tangential component of the acceleration of P_3, A_{P3}^t, is normal to $O_3 P_3$, but is otherwise unknown. Its construction will be temporarily postponed. The con-

$\omega_3 = 6.2/4.5 = 1.4$ rad/s² (CCW)

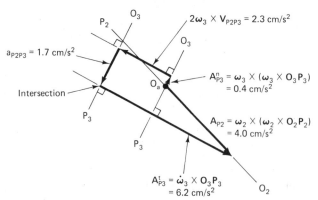

P_2 O_3 $2\omega_3 \times V_{P2P3} = 2.3$ cm/s²

O_3

$a_{P2P3} = 1.7$ cm/s²

Intersection

O_a

$A_{P3}^n = \omega_3 \times (\omega_3 \times O_3 P_3)$
    $= 0.4$ cm/s²

P_3

$A_{P2} = \omega_2 \times (\omega_2 \times O_2 P_2)$
    $= 4.0$ cm/s²

P_3

$A_{P3}^t = \dot{\omega}_3 \times O_3 P_3$
    $= 6.2$ cm/s²

O_2

Figure 2.17   Acceleration Diagram

struction of $\mathbf{a}_{P2P3}$, which is known to be parallel to O_3P_3, will also be postponed because it cannot be completely specified. These vectors will be drawn on the acceleration diagram after all the known vectors are in place. Postponing their construction permits the graphical solution to be obtained by the intersection without further adjustments to the diagram.

The next term to be constructed is the Coriolis acceleration, $2\boldsymbol{\omega}_3 \times \mathbf{v}_{P2P3}$. This term is expected any time the velocity analysis yields a relative velocity in a rotating reference frame. Referring to the velocity diagram and using the right-hand rule show that this acceleration is normal to the construction line O_3P_3, pointing upward to the left, perpendicular to $\mathbf{A}_{P3}^n$, as shown in Figure 2.17.

At this stage of the construction, all of the known acceleration vectors have been drawn on the diagram. Only the postponed accelerations remain to be drawn and they can be added in any order. Let's take $\mathbf{a}_{P2P3}$ first; it is parallel to O_3P_3, which makes it perpendicular to $2\boldsymbol{\omega}_3 \times \mathbf{v}_{P2P3}$. Figure 2.17 shows a construction line drawn through the head of the Coriolis vector parallel to O_3P_3. To complete the diagram, $\mathbf{A}_{P3}^t$ must be added to $\mathbf{a}_{P2P3}$ so that it reaches the head of $\mathbf{A}_{P2}$. It is normal to O_3P_3, which makes it perpendicular to $\mathbf{a}_{P2P3}$. The diagram is closed by constructing a line normal to O_3P_3 through the head of $\mathbf{A}_{P2}$. The intersection of this construction line with the previous one is the simultaneous solution for the two unknown accelerations.

The student should be satisfied that the intuitive feel for the sense of the relative and angular accelerations is in agreement with the graphical results. The reconciliation of preconceived notions of the size and sense of the accelerations with the graphical results can be a very enlightening exercise. Guessing the sense of a velocity is usually much easier than guessing the sense of an acceleration.

2.6 SLIDING CONTACT: CURVED RELATIVE MOTION

Figure 2.18 shows a mechanism that is very similar to the one shown in Figure 2.15. The important difference is the path that pin P_2 follows relative to the coordinate system attached to member (3) at P_3. If the xy coordinates are assigned the angular velocity and acceleration of member (3), the pin will follow a curved path that is fixed in this moving system. At the instant under consideration, the pin happens to be at the origin of this system, but it is traveling on a curve whose radius of curvature, ρ_3, is the distance between C_3 and P_3. The origin of this radius is at C_3, a point fixed in member (3).

The velocity analysis of this mechanism is the same as for the Geneva wheel. The curvature of the relative path has no effect on the formulation. Figure 2.19 is the velocity diagram for the mechanism. Velocity $\mathbf{V}_{P2}$ is normal to the construction line O_2P_2. Velocity $\mathbf{V}_{P3}$ is normal to the construction line O_3P_3. Velocity $\mathbf{v}_{P2P3}$ is normal to the radius of curvature, i.e., tangent to the curved path at P_3. Closure of the velocity diagram is achieved by intersecting the construction lines for $\mathbf{V}_{P3}$ and $\mathbf{v}_{P2P3}$.

The relative acceleration of pin P_2 with respect to its curved path fixed in the

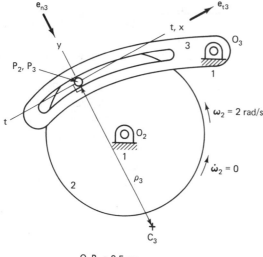

$O_2P_2 = 3.5$ cm
$\rho_3 = C_3P_3 = 8.0$ cm
$O_3P_3 = 7.0$ cm

Figure 2.18   Sliding Contact, Curved Relative Motion

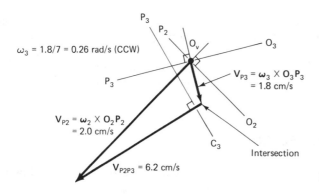

Figure 2.19   Velocity Diagram

xy coordinate system consists of two components in this case. One is tangent to the curve as before, and the other is normal to the curve directed along the radius of curvature from P_3 toward C_3. The magnitude and sense of the tangential acceleration $\mathbf{a}_{P2P3}^t$ is unknown. The normal acceleration $\mathbf{a}_{P2P3}^n$ is known. In path coordinates, it is

$$\mathbf{a}_{P2P3}^n = (|V_{P2P3}|^2/\rho_3)\mathbf{e}_{n3}$$

The acceleration of the pin is given by

$$\mathbf{A}_{P2} = \mathbf{A}_{P3} + \mathbf{a}_{P2P3}^t + \mathbf{a}_{P2P3}^n + 2\boldsymbol{\omega}_3 \times \mathbf{v}_{P2P3}$$

The acceleration $\mathbf{A}_{P3}$ is the acceleration of the origin of the xy coordinate system. If a coordinate system is fixed to member (3) at O_3, this acceleration is given by

$$\mathbf{A}_{P3} = \mathbf{A}_{P3}^n + \mathbf{A}_{P3}^t = \boldsymbol{\omega}_3 \times (\boldsymbol{\omega}_3 \times \mathbf{O}_3\mathbf{P}_3) + \dot{\boldsymbol{\omega}}_3 \times \mathbf{O}_3\mathbf{P}_3$$

The first term of this equation is the centripetal, or normal, acceleration, which is directed from P_3 toward O_3 along $P_3 O_3$. The magnitude of this vector can be found from the results of the velocity analysis. The second term is the tangential acceleration, which is normal to $P_3 O_3$. Its sense and size are unknown.

The acceleration of P_2 is determined in a similar way. Since $\dot{\omega}_2 = \mathbf{O}$, its formulation contains only a normal component, i.e.,

$$\mathbf{A}_{P2} = \boldsymbol{\omega}_2 \times (\boldsymbol{\omega}_2 \times \mathbf{O}_2 \mathbf{P}_2)$$

The unknown vectors in this analysis are the two tangential components, i.e., the relative acceleration $\mathbf{a}_{P2P3}^t$ and $\mathbf{A}_{P3}^t = \dot{\omega}_3 \times \mathbf{O}_3 \mathbf{P}_3$. These will be the last vectors drawn on the acceleration diagram. Their intersection will yield the simultaneous solution for the angular acceleration $\dot{\omega}_3$ and $\mathbf{a}_{P2P3}^t$.

Figure 2.20 shows the acceleration diagram. The acceleration of the pin along $P_2 O_2$ is constructed first.

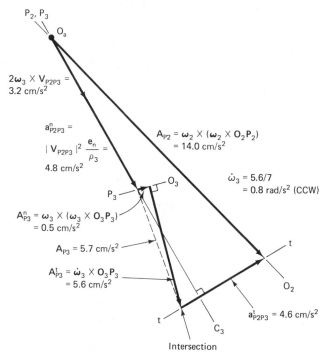

P_2, P_3

O_a

$2\boldsymbol{\omega}_3 \times \mathbf{V}_{P2P3} =$
3.2 cm/s^2

$a_{P2P3}^n =$

$|\mathbf{V}_{P2P3}|^2 \dfrac{\mathbf{e}_n}{\rho_3} =$

4.8 cm/s^2

$\mathbf{A}_{P2} = \boldsymbol{\omega}_2 \times (\boldsymbol{\omega}_2 \times \mathbf{O}_2 \mathbf{P}_2)$
$= 14.0 \text{ cm/s}^2$

O_3

$\dot{\omega}_3 = 5.6/7$
$= 0.8 \text{ rad/s}^2 \text{ (CCW)}$

P_3

$A_{P3}^n = \boldsymbol{\omega}_3 \times (\boldsymbol{\omega}_3 \times \mathbf{O}_3 \mathbf{P}_3)$
$= 0.5 \text{ cm/s}^2$

$\mathbf{A}_{P3} = 5.7 \text{ cm/s}^2$

t

$\mathbf{A}_{P3}^t = \dot{\omega}_3 \times \mathbf{O}_3 \mathbf{P}_3$
$= 5.6 \text{ cm/s}^2$

O_2

t

$a_{P2P3}^t = 4.6 \text{ cm/s}^2$

C_3

Intersection

Figure 2.20   Acceleration Diagram

From the origin of the diagram, O_a, for the terms on the right-hand side of the equation for $\mathbf{A}_{P2}$, the Coriolis acceleration $2\boldsymbol{\omega}_3 \times \mathbf{v}_{P2P3}$ is drawn first along the construction line $P_3 C_3$. The sense of this acceleration is obtained from the velocity diagram using the right-hand rule. Next, the normal component of the relative acceleration $\mathbf{a}_{P2P3}^n$ is added along the same construction line. Finally, the normal component of the acceleration of P_3 is drawn along $P_3 O_3$. This completes the drawing of the known vectors.

Two construction lines are now drawn, one normal to $P_3 O_3$ through the head of $\mathbf{A}_{P3}^n$ and one normal to $P_3 C_3$ through the head of $\mathbf{A}_{P2}$. The intersection of these lines determines the remaining unknown vectors, $\mathbf{A}_{P3}^t$ and $\mathbf{a}_{P2P3}^t$, as shown in Figure 2.20. Acceleration $\mathbf{a}_{P2P3}^t$ is tangent to the curved path in member (3) at P_3, i.e., along line t–t. The total acceleration of P_3 is the vector sum $\mathbf{A}_{P3}^t + \mathbf{A}_{P3}^n$ shown in Figure 2.20 as a dashed line.

A simpler mechanism that has the same instantaneous kinematics as the original mechanism of Figure 2.18 is shown in Figure 2.21. This is called the "equivalent mechanism" and its existence stems from the concept of the radius of curvature.

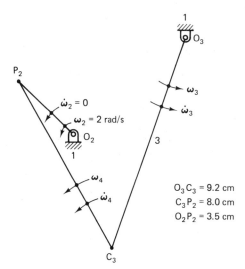

Figure 2.21 Equivalent Mechanism

Recall that the arc drawn by the radius of curvature conforms to the path to the order of the second derivative [Equation (2.9)]. That means that the original mechanism's instantaneous kinematics would not be changed if the radius of curvature was replaced by a rigid link joining C_3 and P_2. Link $O_3 C_3$ shown in Figure 2.21 is merely an extension of member (3) to include the point C_3, which is fixed in that member.

The newly added link, $P_2 C_3$, has an instantaneous angular velocity ω_4 that will be used in the acceleration analysis. There is no member in the original mechanism that has this angular velocity.

The velocity analysis of this equivalent four-bar mechanism is shown in Figure 2.22. Velocity $\mathbf{V}_{P3}$ is also shown. A comparison of Figures 2.22 and 2.19 will show that triangle $O_v P_2 P_3$ of Figure 2.22 is the velocity diagram drawn in Figure 2.19. A vector joining P_3 and P_2 would yield $\mathbf{v}_{P2P3}$ just as it appears in Figure 2.19. The acceleration diagram for the equivalent mechanism is shown in Figure 2.23.

A formal proof of equivalency can be obtained using the kinematic equations. The equation that mathematically expresses the graphical velocity solution of Figure 2.19 is

$$\mathbf{V}_{P2} = \mathbf{V}_{P3} + \mathbf{v}_{P2P3}$$

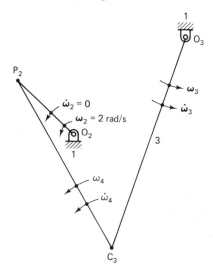

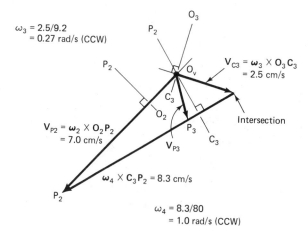

Figure 2.22 Velocity Diagram of Equivalent Mechanism

The relative-velocity term can be expressed in terms of a vector joining C_3 and P_2 and its angular velocity relative to the auxiliary coordinate xy fixed to member (3), $\boldsymbol{\omega}_{43}$. In these terms, $\mathbf{v}_{P2P3} = \boldsymbol{\omega}_{43} \times \mathbf{C}_3\mathbf{P}_2$. The velocity of C_3 can be introduced by noting that

$$\mathbf{V}_{P3} = \mathbf{V}_{C3} + \boldsymbol{\omega}_3 \times \mathbf{C}_3\mathbf{P}_3 = \mathbf{V}_{C3} + \boldsymbol{\omega}_3 \times \mathbf{C}_3\mathbf{P}_2$$

since $C_3P_3 = C_3P_2$.

Substituting these into the velocity equation yields

$$\mathbf{V}_{P2} = \mathbf{V}_{C3} + \boldsymbol{\omega}_{43} \times \mathbf{C}_3\mathbf{P}_2 + \boldsymbol{\omega}_3 \times \mathbf{C}_3\mathbf{P}_2 = \mathbf{V}_{C3} + (\boldsymbol{\omega}_3 + \boldsymbol{\omega}_{43}) \times \mathbf{C}_3\mathbf{P}_2$$

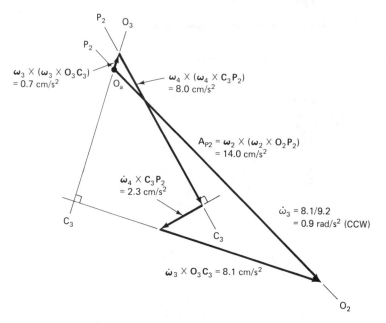

Figure 2.23   Acceleration Diagram of Equivalent Mechanism

Since $\omega_4 = \omega_3 + \omega_{43}$, this equation can be reduced to $V_{P2} = V_{C3} + \omega_4 \times C_3 P_2$. This is the velocity equation for link (4) of the equivalent mechanism in Figure 2.21.

The equivalency of accelerations can be shown in a similar manner. The graphical acceleration solution of Figure 2.20 is based on the formulation

$$A_{P2} = A_{P3} + a^n_{P2P3} + a^t_{P2P3} + 2\omega_3 \times v_{P2P3}$$

The substitutions in this case are

$$A_{P3} = A_{C3} + a^n_{P3C3} + a^t_{P3C3} = A_{C3} + \omega_3 \times (\omega_3 \times C_3 P_2) + \dot{\omega}_3 \times C_3 P_2$$

$$a^n_{P2P3} = \omega_{43} \times (\omega_{43} \times C_3 P_2)$$

$$a^t_{P2P3} = \dot{\omega}_{43} \times C_3 P_2$$

$$2\omega_3 \times v_{P2P3} = 2\omega_3 \times (\omega_{43} \times C_3 P_2)$$

With these, the acceleration equation becomes

$$A_{P2} = A_{C3} + \omega_3 \times (\omega_3 \times C_3 P_2) + \dot{\omega}_3 \times C_3 P_2 + \omega_{43} \times (\omega_{43} \times C_3 P_2)$$
$$+ \dot{\omega}_{43} \times C_3 P_2 + 2\omega_3 \times (\omega_{43} \times C_3 P_2)$$

The terms containing angular accelerations combine to form

$$\dot{\omega}_3 \times C_3 P_2 + \dot{\omega}_{43} \times C_3 P_2 = \dot{\omega}_4 \times C_3 P_2$$

which is the transverse acceleration of the "equivalent link" (4).

Substituting $\omega_{43} = \omega_4 - \omega_3$ into the remaining terms reduces their sum to $\omega_4 \times (\omega_4 \times C_3 P_2)$.

The acceleration equation then becomes

$$\mathbf{A}_{P2} = \mathbf{A}_{C3} + \omega_4 \times (\omega_4 \times \mathbf{C}_3 \mathbf{P}_2) + \dot{\omega}_4 \times \mathbf{C}_3 \mathbf{P}_2$$

which is the acceleration equation for the equivalent mechanism.

A word about accuracy is appropriate at this point. Obviously, the accuracy of a graphical solution depends on the accuracy with which it is drawn. Large drawings tend to be more accurate. Mechanisms with fewer elements, such as equivalent mechanisms, should be more accurate. (Note the slight differences in the results obtained in Figures 2.19, 2.20, 2.22, and 2.23). The utility of the graphical method is speed. With a little practice, one can quickly draw diagrams of the kinematics, which also yield reasonably accurate answers.

2.7 ROLLING CONTACT WITHOUT SLIP

The matching of the kinematics across a rolling contact interface presents some special problems. The following example illustrates what they are. Figure 2.24 shows a cylinder rolling on a stationary flat surface. The center of the cylinder, C_2, is translating horizontally with velocity and acceleration, $\mathbf{V}_{C2}$ and $\mathbf{A}_{C2}$, respectively. Contact points P_2 and P_1 are coincident at this instant. Since P_1 is at rest, P_2 has the same velocity as P_1 if there is no slip, i.e., $\mathbf{V}_{P1} = \mathbf{V}_{P2} = 0$.

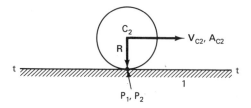

Figure 2.24   Rolling Cylinder

The same cannot be said for the accelerations $\mathbf{A}_{P1}$ and $\mathbf{A}_{P2}$. Although $\mathbf{A}_{P1}$ is zero, $\mathbf{A}_{P2}$ is not. The transverse component of $\mathbf{A}_{P2}$ along the line t–t is zero when there is no slip, but the normal component of $\mathbf{A}_{P2}$ is not. The acceleration of $\mathbf{A}_{P2}$ is given by the equation

$$\mathbf{A}_{P2} = \mathbf{A}_{C2} + \mathbf{A}_{P2C2}^t + \mathbf{A}_{P2C2}^n = \mathbf{A}_{C2} + \dot{\omega}_2 \times \mathbf{R} + \omega_2 \times (\omega_2 \times \mathbf{R})$$

The components of $\mathbf{A}_{P2}$ along t–t are $\mathbf{A}_{C2} + \dot{\omega}_2 \times \mathbf{R}$ and they must add to zero. This leaves $\mathbf{A}_{P2}$ equal to its normal component $\omega_2 \times (\omega_2 \times \mathbf{R})$, which is directed normal to t–t toward C_2.

From this example, we learn that the transverse components of the accelerations $\mathbf{A}_{P2}$ and $\mathbf{A}_{P1}$ are equal, but its normal components are not. This rule applies to all rolling contact points that do not slip.

The example of Figure 2.25 illustrates the use of this matching condition. In this cam mechanism, it is assumed that the center of disk (4) rotates with a circular motion around the fixed point C_1 while rolling on a circular surface of radius r_1.

Since point P_4 is in contact with point P_1, which has a zero velocity, the velocity of P_4 is also zero. The velocities of link (3) and disk (4) can be found by treating the mechanism as though it were a four-bar linkage with P_4 attached to ground by a pin. The dashed line, $B_4 P_4$, has the angular velocity of disk (4), which it replaces. The velocity diagram for this mechanism is shown in Figure 2.26.

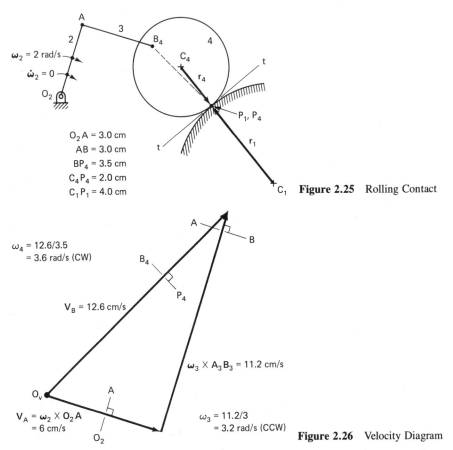

Figure 2.25   Rolling Contact

Figure 2.26   Velocity Diagram

The acceleration analysis cannot be performed so simply because the acceleration of P_4 is not known. If disk (4) were rolling on a straight surface along t–t, the acceleration of P_4 would be $\boldsymbol{\omega}_4 \times (\boldsymbol{\omega}_4 \times \mathbf{r}_4)$, as found in the previous example. In that example, the center of the disk did not have an acceleration component normal to the tangent line t–t. In this case, it does due to the circular motion of C_4 about C_1.

It will be necessary to find the acceleration of P_4 before proceeding to solve the original problem. That component contributes to the acceleration of C_4. Figure 2.27 shows disk (4) and its mating member joined by a new link (5) that has been added

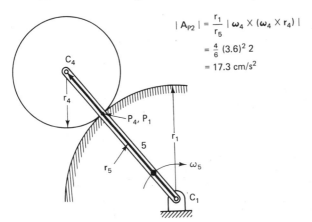

$$| A_{P2} | = \frac{r_1}{r_5} | \boldsymbol{\omega}_4 \times (\boldsymbol{\omega}_4 \times r_4) |$$

$$= \frac{4}{6} (3.6)^2 \, 2$$

$$= 17.3 \text{ cm/s}^2$$

Figure 2.27   Acceleration of P_4

to facilitate the analysis. This new link does not change the kinematics of the mechanism, although it might be a useful addition if it is found that points P_4 and P_1 tend to separate.

The velocity of C_4 is given by

$$\mathbf{V}_{C4} = \boldsymbol{\omega}_5 \times \mathbf{r}_5$$

The velocity of P_4 is given by

$$\mathbf{V}_{P4} = \mathbf{V}_{C4} + \boldsymbol{\omega}_4 \times \mathbf{r}_4 = \boldsymbol{\omega}_5 \times \mathbf{r}_5 + \boldsymbol{\omega}_4 \times \mathbf{r}_4 = \mathbf{0}$$

The angular velocity of the added link is found from

$$\boldsymbol{\omega}_4 \times \mathbf{r}_4 = -\boldsymbol{\omega}_5 \times \mathbf{r}_5$$

Since $\mathbf{r}_4$ and $\mathbf{r}_5$ are in opposite directions, $\boldsymbol{\omega}_4$ and $\boldsymbol{\omega}_5$ are in the same direction. The magnitudes of the angular velocities scale according to $\boldsymbol{\omega}_5 = (r_4/r_5)\boldsymbol{\omega}_4$, where $\boldsymbol{\omega}_4$ is known from the velocity analysis.

The acceleration of C_4 is given by

$$\mathbf{A}_{C4} = \mathbf{A}_{C4}^n + \mathbf{A}_{C4}^t = \boldsymbol{\omega}_5 \times (\boldsymbol{\omega}_5 \times \mathbf{r}_5) + \dot{\boldsymbol{\omega}}_5 \times \mathbf{r}_5$$

The acceleration of P_4 is

$$\mathbf{A}_{P4} = \mathbf{A}_{C4} + \mathbf{A}_{P4C4}^n + \mathbf{A}_{P4C4}^t = \boldsymbol{\omega}_5 \times (\boldsymbol{\omega}_5 \times \mathbf{r}_5) + \dot{\boldsymbol{\omega}}_5 \times \mathbf{r}_5$$

$$+ \boldsymbol{\omega}_4 \times (\boldsymbol{\omega}_4 \times \mathbf{r}_4) + \dot{\boldsymbol{\omega}}_4 \times \mathbf{r}_4$$

The transverse components of the accelerations $\mathbf{A}_{P1}$ and $\mathbf{A}_{P4}$ are equal. Since the transverse component of the acceleration of P_4 is zero,

$$\mathbf{A}_{P4}^t = \dot{\boldsymbol{\omega}}_5 \times \mathbf{r}_5 + \dot{\boldsymbol{\omega}}_4 \times \mathbf{r}_4 = \mathbf{0}$$

This equation will yield the angular acceleration of the added link (5) once the angular acceleration of member (4) is found.

The normal component of the acceleration of P_4 is

$$\mathbf{A}_{P4}^n = \boldsymbol{\omega}_5 \times (\boldsymbol{\omega}_5 \times \mathbf{r}_5) + \boldsymbol{\omega}_4 \times (\boldsymbol{\omega}_4 \times \mathbf{r}_4)$$

By using the results of the velocity analysis, this can be rewritten as

$$\mathbf{A}_{P4} = \mathbf{A}_{P4}^n = (r_1/r_5)[\boldsymbol{\omega}_4 \times (\boldsymbol{\omega}_4 \times \mathbf{r}_4)]$$

The scalar ratio $r_1/r_5 = r_1/(r_1 + r_4)$ accounts for the finite radius of curvature of fixed member (1). This ratio approaches unity as r_1 approaches infinity, yielding in the limit the result obtained earlier for rolling on a straight surface.

The acceleration analysis of the mechanism can now proceed using the four-bar mechanism of the velocity analysis.

Figure 2.28 shows the acceleration diagram for the mechanism. Closure at B_4 is obtained by constructing $\dot{\boldsymbol{\omega}}_3 \times \mathbf{AB}$ normal to $\boldsymbol{\omega}_3 \times (\boldsymbol{\omega}_3 \times \mathbf{AB})$ and $\dot{\boldsymbol{\omega}}_4 \times \mathbf{P}_4\mathbf{B}_4$ normal to $\boldsymbol{\omega}_4 \times (\boldsymbol{\omega}_4 \times \mathbf{P}_4\mathbf{B}_4)$ after the known vectors $\mathbf{A}_A$, $\boldsymbol{\omega}_3 \times (\boldsymbol{\omega}_3 \times \mathbf{AB})$, $\mathbf{A}_{P4}$, and $\boldsymbol{\omega}_4 \times (\boldsymbol{\omega}_4 \times \mathbf{P}_4\mathbf{B}_4)$ are drawn. Note that the acceleration $\mathbf{A}_{P4}^n$ is directed toward the center of curvature C_4.

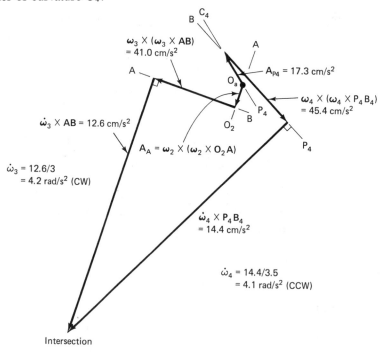

Figure 2.28   Acceleration Diagram

2.8 EQUIVALENT MECHANISMS

The concept of an equivalent mechanism was introduced in several of the preceding examples. The substitution of an equivalent mechanism for the original substantially reduced the complexity of the kinematic analysis. By realizing this, considerable effort has been devoted to the discovery and classification of equivalent mechanisms for most of the commonly encountered engineering mechanisms. Figure 2.29 is a compilation of these results.

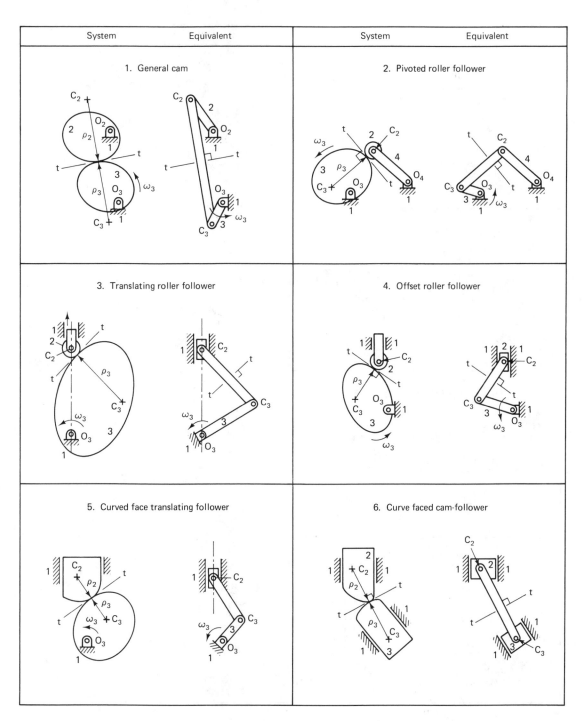

Figure 2.29 Equivalent Mechanisms

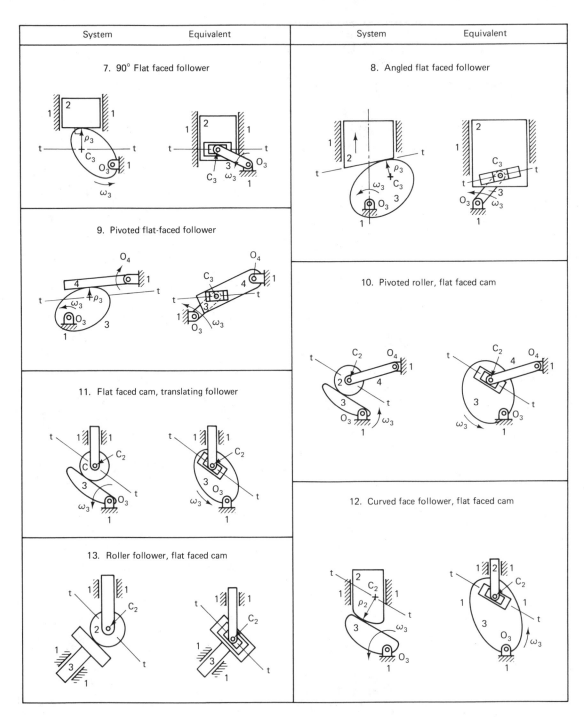

Figure 2.29 Equivalent Mechanisms (*continued*)

Cases 1 through 6 are based on the concept that the radii of curvature of the contacting surfaces remain constant during contact. In each of these cases, the centers of curvature are instantaneously joined by a rigid link. Since the centers of curvature are fixed in their respective members, they are joined to ground by replacement links containing these centers as fixed points. The previous example was an application of case 2.

In cases 7 through 12, one of the contacting surfaces is a straight line with an infinite radius of curvature. The equivalent mechanism for each of these cases can be found by selecting its antecedent from among the first six cases and then letting one radius of curvature approach infinity.

As an example, case 5 is the antecedent of case 8, i.e., case 8 is obtained from case 5 by letting ρ_2 approach infinity. The velocity diagram for the equivalent mechanism of case 5 is shown in Figure 2.30(a). The original mechanisms of cases 5 and 8 differ in the size of ρ_2, which is finite in the first case and infinite in the second. To find the equivalent mechanism for case 8, we must determine the effect of letting ρ_2 approach infinity in case 5. Since $\mathbf{V}_{C3} = \boldsymbol{\omega}_3 \times \mathbf{O}_3\mathbf{C}_3$ is a known vector, it is constructed first in Figure 2.30(a). The line of action of $\mathbf{V}_{C2}$ is also known, i.e., up or down, so a construction line in this direction can be drawn. Figure 2.30(a) is closed by constructing a vector normal to the radii of curvature. In case 5, this closure vector is called $\boldsymbol{\omega}_{C2C3} \times \mathbf{C}_3\mathbf{C}_2$ for obvious reasons.

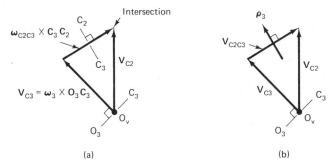

Figure 2.30 Velocity Diagrams for (a) Case 5 and (b) Case 8

The velocity diagram for the original mechanism of case 8 would begin by drawing $\mathbf{V}_{C3}$ and the vertical construction line for $\mathbf{V}_{C2}$, Figure 2.30(b). The two cases have these constructions in common. Closure is made once again by drawing a vector normal to the radii of curvature. The fact that one of the radii is infinite should not affect the way this vector is constructed, although it will affect the name that it is given. Obviously, it cannot be called $\boldsymbol{\omega}_{C2C3} \times \mathbf{C}_3\mathbf{C}_2$ anymore. It can be designated as $\mathbf{V}_{C2C3}$ since it joins $\mathbf{V}_{C3}$ to $\mathbf{V}_{C2}$. It has the same length and sense as $\boldsymbol{\omega}_{C2C3} \times \mathbf{C}_3\mathbf{C}_2$, but a different interpretation is required. The equivalent mechanism shown for case 8 has the velocity diagram of Figure 2.30(b).

So far we have verified that the velocities of members (2) and (3) are described by the equivalent mechanism shown in case 8. Before proceeding to show that their accelerations are also correctly predicted by this mechanism, it should be noted that the vector equivalent to the product $\boldsymbol{\omega}_{C2C3} \times \mathbf{C}_3\mathbf{C}_2$ remains finite as we move from case 5 to case 8. It just changes its name. From this, we conclude that $\boldsymbol{\omega}_{C2C3}$ approaches zero as $\mathbf{C}_3\mathbf{C}_2$ approaches infinity, maintaining the product $\boldsymbol{\omega}_{C2C3} \times \mathbf{C}_3\mathbf{C}_2$ as a finite nonzero quantity. This observation will prove to be useful in the analysis of the accelerations.

The acceleration diagram for the equivalent mechanism of case 5 is shown in Figure 2.31(a) for $\dot{\boldsymbol{\omega}}_3 = \mathbf{O}$.

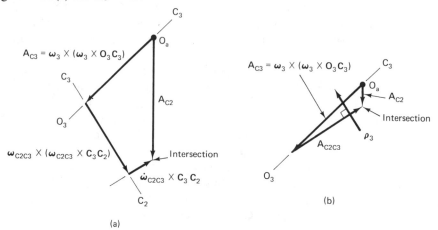

Figure 2.31 Acceleration Diagrams for (a) Case 5 and (b) Case 8

The acceleration $\mathbf{A}_{C3}$ is known initially. The line of action of $\mathbf{A}_{C2}$ is also known, i.e., up or down. The normal acceleration $\boldsymbol{\omega}_{C2C3} \times (\boldsymbol{\omega}_{C2C3} \times \mathbf{C}_3\mathbf{C}_2)$ can be determined from the results of the velocity analysis. The diagram is closed by intersecting the vertical construction line for $\mathbf{A}_{C2}$ with $\dot{\boldsymbol{\omega}}_{C2C3} \times \mathbf{C}_3\mathbf{C}_2$ drawn from the end of $\boldsymbol{\omega}_{C2C3} \times (\boldsymbol{\omega}_{C2C3} \times \mathbf{C}_3\mathbf{C}_3)$.

As we move from case 5 to case 8, the term $\boldsymbol{\omega}_{C2C3} \times (\boldsymbol{\omega}_{C2C3} \times \mathbf{C}_3\mathbf{C}_2)$ vanishes because $\boldsymbol{\omega}_{C2C3}$ approaches zero while $\boldsymbol{\omega}_{C2C3} \times \mathbf{C}_3\mathbf{C}_2$ remains finite. Closure is now obtained by intersecting the vertical construction line for $\mathbf{A}_{C2}$ with a construction line drawn normal to ρ_3 through the end of $\mathbf{A}_{C3}$, as shown in Figure 2.31(b). This is the acceleration diagram for the equivalent mechanism shown for case 8. Notice that the total relative acceleration $\mathbf{A}_{C2C3}$ is tangent to the path, i.e., along t–t. Since it has no component normal to the path, a straight slot is shown in the extension of member (2).

We have shown that the equivalent mechanism for case 8 is the limiting version of the one for case 5 with regard to both velocity and acceleration. A completely equivalent mechanism must satisfy both criteria. Similar procedures can be used to verify the other equivalent mechanism for cases where one radius of curvature is infinite. Actually, there are many equivalent mechanisms that can be devised to replace the original. The ones shown in cases 1 to 13 in Figure 2.29 are the simplest of these.

Figure 2.32(a) shows a simple cam-follower mechanism made by eccentrically mounting a cylinder. Figure 2.32(b) shows the equivalent mechanism. Since all

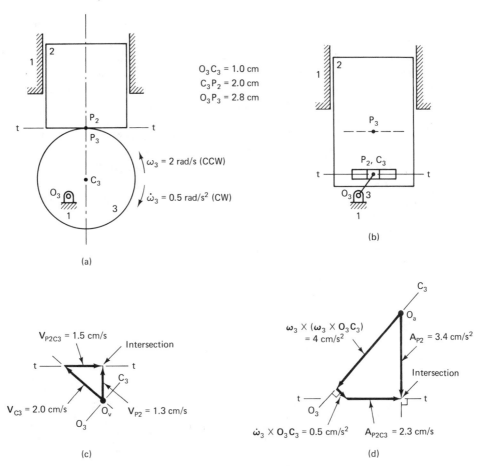

Figure 2.32 (a) Cam Mechanism, (b) Equivalent Mechanism, (c) Velocity Diagram, (d) Acceleration Diagram

points fixed in member (2) have the same velocity and acceleration, point P_2 is shown coincident with C_3 at this instant; line t–t has also been transferred with P_2. Figures 2.32(c) and (d) are the velocity and acceleration diagrams, respectively, for the equivalent mechanism. At this instant, the only important geometric properties of the cam are its radius of curvature, $P_3 C_3$, and the location of the center of curvature. The shape of the rest of the cam, circular or not, is unimportant once these quantities are known.

2.9 COMPLEX-VARIABLE METHOD

In the previous sections, the solutions of the vector kinematic equations have always been obtained using graphical methods. These methods yield quick visual solutions that are as accurate as graphical methods permit. Unfortunately, a complete history of the kinematics of a mechanism can only be obtained at the expense of drawing many such diagrams. Graphical solutions do not readily lend themselves to the discovery of specific features of the kinematics such as the configurations that yield the largest component velocities or accelerations. A search for these conditions clearly calls for an analytical formulation. Highly accurate results also require greater precision than can be attained via graphical construction.

In this section, an analytical vector method using complex numbers will be introduced that overcomes these deficiencies. The concepts developed here will also be used in later sections devoted to balancing.

The basic features of the complex variable can be illustrated using the Argand diagram, or complex plane, shown in Figure 2.33. The vector **r** shown on this diagram would ordinarily be represented in terms of unit vectors i and j and projections x and y. The position vector would then be written

$$\mathbf{r} = x\mathbf{i} + y\mathbf{j}$$

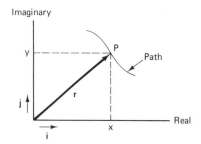

Figure 2.33   Position Vector in Complex Plane

Using two unit vectors to identify the components is really redundant since one alone would suffice if it were conventional to associate it with a particular projection. For example, a component could be identified using a single marker, say, i, for the y component. With this convention, the vector could then be written $\mathbf{r} = x + iy$ provided it is understood that the coefficient of i is the y component and the other quantity is the x component.

Carrying this idea a step further, we would also want to establish an "algebra" for this notation that would permit certain operations to be performed, such as addition and differentiation. We are indeed fortunate in this regard because if we chose the marker i to be the imaginary quantity $\sqrt{-1}$, a very simple algebra can be devised that makes these operations easy to perform. The labeling of the Argand diagram coordinate x as "real" and y as "imaginary" is in this spirit.

The development that follows will show that i not only identifies components, but also functions as an operator. We will find that it will not be necessary to know the "value" of i, only its operational properties. Figure 2.34 illustrates the algebra of addition using complex variables, where XY are the primary fixed coordinates and xy are the secondary moving auxiliary coordinates. The sum of the vectors **R** and **r** is

$$\mathbf{R} + \mathbf{r} = (X + x) + (Y + y)i$$

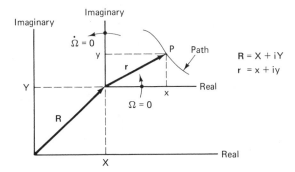

Figure 2.34 Algebra of Addition for Complex Numbers

In previous applications, it was not necessary to align these two coordinate systems. The simple addition rule used is correct only if they *are* aligned. In the past, the xy coordinates have also been allowed to rotate. Many of the advantages of the complex-variable method are sacrificed if this is allowed. For this reason, a nonrotating aligned coordinate system will always be used in the applications covered in this text. Note that no constraint is placed on the translation of xy relative to XY.

The original complex-variable formulation of the vector locating point P is cartesian. It is often much more convenient to perform operations using a polar complex-variable formulation. Figure 2.35 shows that this can be done by introducing the length of the vector $\mathbf{r}$ (its modulus) and the angle θ the vector makes with the real axis (its argument) measured in the positive sense. The result is

$$\mathbf{r} = x + iy = r\,(\cos\theta + i\sin\theta)$$

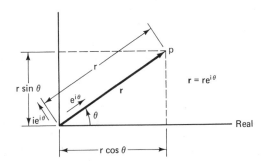

Figure 2.35 Polar Form of Position Vector

With the aid of the following identities,

$$\sin\theta = (e^{i\theta} - e^{-i\theta})/2i \qquad \cos\theta = (e^{i\theta} + e^{-i\theta})/2$$

this trigonometric form reduces to the very compact exponential form

$$\mathbf{r} = re^{i\theta} \tag{2.10}$$

When written this way, the right side is the product of the scalar length of the vector $\mathbf{r}$ and a unit vector $e^{i\theta}$ in the direction of $\mathbf{r}$, as shown in Fig. 2.35.

The differentiation of this vector proceeds in a straightforward manner, i.e.,

$$d\mathbf{r}/dt = \dot{\mathbf{r}} = \dot{r}e^{i\theta} + ri\dot{\theta}e^{i\theta}$$

The first term in this equation represents the radial growth rate of the position vector, or the radial velocity component of the point P. The meaning of the second term is not so obvious since i appears separated from the exponential.

It is once again convenient to separate the product into two parts, r and $i\dot{\theta}e^{i\theta}$. The latter is the derivative of the original unit vector $e^{i\theta}$, i.e.,

$$d/dt\,(e^{i\theta}) = i\dot{\theta}e^{i\theta}$$

Before trying to interpret this term, recall that $e^{i\theta} = \cos\theta + i\sin\theta$. If θ is chosen to be $\pi/2$,[†] then $i = e^{i\pi/2}$. The derivative now becomes

$$d/dt\,(e^{i\theta}) = \dot{\theta}e^{i(\theta+\pi/2)}$$

The quantity $ie^{i\theta} = e^{i(\theta+\pi/2)}$ is a new unit vector advanced 90° ahead of the original radial unit vector $e^{i\theta}$. Figure 2.35 shows the relationship between these two vectors.

This new vector, created by differentiation, is a unit vector in the tangential direction. Its scalar coefficient, $r\dot{\theta}$, is the magnitude of the tangential velocity, so taken together, the second term of the derivative is the tangential velocity of point P.

The straightforward differentiation of the complex-variable form of the position vector has generated the polar-coordinate form of the velocity equation. In this formulation, the unit vectors in the radial and tangential directions are given by $e^{i\theta}$ and $ie^{i\theta} = e^{i(\theta+\pi/2)}$. Another differentiation would yield the polar form of the acceleration equation.

Because of its differentiation properties, the complex-variable position vector is aptly suited to the kinematic analysis of two-dimensional mechanisms. The following example illustrates how the complex variable may be used to obtain analytical expressions for the kinematics of such mechanisms.

Figure 2.36 shows a Scotch yoke, which is frequently used as a device to convert rotary motion to translated motion. The tee-shaped member is driven back and forth by pin P_2 on the rotating disk. Figure 2.37 shows the vectors used to describe the mechanism. The vectors used are

> $\mathbf{r}_{P2}$ = a vector of fixed length at 45° rotating with disk (2), locating pin P_2 attached to the disk.
>
> $\mathbf{r}_{P3}$ = a nonrotating vector of variable length at $\theta_4 = 0$, joining point P_3 [fixed on member (3)] to ground.
>
> $\mathbf{r}_{P2P3}$ = a nonrotating vector of variable length at $\theta_3 = 90°$, joining sliding pin P_2 to a point fixed to the slot at P_3.

From the geometry of Figure 2.37,

$$\mathbf{r}_{P3} + \mathbf{r}_{P2P3} = \mathbf{r}_{P2}$$

or

$$r_{P3}e^{i\theta_4} + r_{P2P3}e^{i\theta_3} = r_{P2}e^{i\theta_2}$$

[†] We will use only the principal values of θ, i.e., $0 \le \theta < 2\pi$.

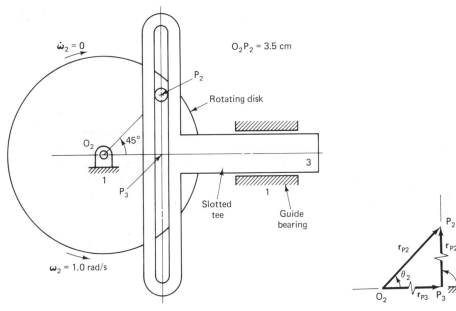

Figure 2.36 Scotch Yoke Mechanism

Figure 2.37 Position Vectors for the Scotch Yoke

In the preceding formula, the following are constants

$$\theta_4 = 0°$$
$$\theta_3 = 90°$$
$$r_{P2} = 3.5 \text{ cm}$$

The variable quantities are r_{P3}, r_{P2P3}, and θ_2. The kinematic inputs are $\dot{\theta}_2 = \omega_2 = 1.0$ rad/s and $\ddot{\theta}_2 = \dot{\omega}_2 = 0$.

Differentiating the geometric expression yields

$$\dot{r}_{P3}e^{i\theta_4} + \dot{r}_{P2P3}e^{i\theta_3} = i\dot{\theta}_2 r_{P2}e^{i\theta_2}$$

This equation can now be rewritten in trigonometric form as

$$\dot{r}_{P3}(\cos\theta_4 + i\sin\theta_4) + \dot{r}_{P2P3}(\cos\theta_3 + i\sin\theta_3) = r_{P2}\dot{\theta}_2(i\cos\theta_2 - \sin\theta_2)$$

By equating the real parts and the imaginary parts, two equations are formed:

$$\dot{r}_{P3}\cos\theta_4 + \dot{r}_{P2P3}\cos\theta_3 = -r_{P2}\dot{\theta}_2\sin\theta_2$$

$$\dot{r}_{P3}\sin\theta_4 + \dot{r}_{P2P3}\sin\theta_3 = r_{P2}\dot{\theta}_2\cos\theta_2$$

The unknowns in these equations are $\dot{r}_{P3}$, which is the translational velocity of the slotted tee, and $\dot{r}_{P2P3}$, the relative velocity of the pin with respect to the slot. The equations for these velocities are

$$\dot{r}_{P3} = -r_{P2}\dot{\theta}_2(\sin\theta_2 \sin\theta_3 + \cos\theta_2 \cos\theta_3)/(\cos\theta_4 \sin\theta_3 - \sin\theta_4 \cos\theta_3)$$

$$= -2.47 \text{ cm/s}$$

$$\dot{r}_{P2P3} = r_{P2}\dot{\theta}_2(\cos\theta_4 \cos\theta_2 + \sin\theta_4 \sin\theta_2)/(\cos\theta_4 \sin\theta_3 - \sin\theta_4 \cos\theta_3)$$

$$= 2.47 \text{ cm/s}$$

The form of these solutions is of particular interest. The right-hand side of each equation contains the imposed velocity $r_{P2}\dot{\theta}_2 = r_{P2}\omega_2$ modified by a coefficient that is purely geometric, i.e., a function of θ_2, θ_3, and θ_4. Since the imposed accelerations are $r_{P2}\ddot{\theta}_2 = r_{P2}\dot{\omega}_2$ and $r_{P2}\dot{\theta}^2 = r_{P2}\omega_2^2$, we might anticipate that the expressions for the accelerations would be linear combinations of these each modified by purely geometric coefficients. The accelerations can be obtained by differentiating any of the velocity expressions. The results are

$$\ddot{r}_{P3} \cos\theta_4 + \ddot{r}_{P2P3} \cos\theta_3 = -r_{P2}\ddot{\theta}_2 \sin\theta_2 - r_{P2}\dot{\theta}_2^2 \cos\theta_2$$

$$\ddot{r}_{P3} \sin\theta_4 + \ddot{r}_{P2P3} \sin\theta_3 = r_{P2}\ddot{\theta}_2 \cos\theta_2 - r_{P2}\dot{\theta}_2^2 \sin\theta_2$$

Solving for the accelerations,

$$\ddot{r}_{P3} = -r_{P2}\ddot{\theta}_2(\sin\theta_2 \sin\theta_3 + \cos\theta_2 \cos\theta_3)/(\cos\theta_4 \sin\theta_3 - \sin\theta_4 \cos\theta_3)$$
$$- r_{P2}\dot{\theta}_2^2(\cos\theta_2 \sin\theta_3 - \sin\theta_2 \cos\theta_3)/(\cos\theta_4 \sin\theta_3 - \sin\theta_4 \cos\theta_3) = 0$$

$$\ddot{r}_{P2P3} = r_{P2}\ddot{\theta}_2(\cos\theta_4 \cos\theta_2 + \sin\theta_4 \sin\theta_2)/(\cos\theta_4 \sin\theta_3 - \sin\theta_4 \cos\theta_3)$$
$$- r_{P2}\dot{\theta}_2^2(\cos\theta_4 \sin\theta_2 - \sin\theta_4 \cos\theta_2)/(\cos\theta_4 \sin\theta_3 - \sin\theta_4 \cos\theta_3)$$

$$= -4.95 \text{ m/s}^2$$

The form of the equations, as anticipated, indicates how each component of the pin's acceleration and the system's geometry contributes to the acceleration of the slotted tee.

On occasions where it is not convenient to form a closed polygon with the position vectors, it may be necessary to use an auxiliary coordinate system. In the case of the Scotch yoke, one might consider placing the origin of the auxiliary system on member (3), coincident with the pin. This choice would require the use of a zero-length vector joining the pin to the slot. This vector has derivatives representing the relative velocity and acceleration of the pin with respect to the slot. The visualization of such a position vector, particularly its orientation, is not easy. This difficulty is removed if the auxiliary system is placed elsewhere, say, at P_3. The position of P_2 is then given by

$$\mathbf{r}_{P2} = \mathbf{r}_{P3} + \mathbf{r}_{P2P3}$$

where $\mathbf{r}_{P3}$ locates the origin of the xy coordinate system, and $\mathbf{r}_{P2P3}$ locates P_2 in the xy coordinate system; see Figure 2.38(a).

This construction is in keeping with the representation shown earlier in Figure

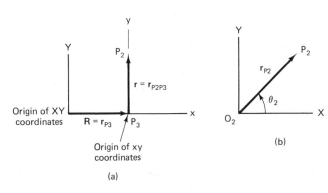

Figure 2.38   Position Vector of P_2 (a) Using an Auxiliary Coordinate System, (b) Using a Fixed Coordinate System

2.1. Since the xy coordinates do not rotate, the vector formulation for the velocity is

$$\mathbf{V} = \dot{\mathbf{R}} + \mathbf{v}_{xy}$$

or

$$\dot{\mathbf{r}}_{P2} = \dot{\mathbf{r}}_{P3} + \dot{\mathbf{r}}_{P2P3} = \dot{r}_{P3}e^{i\theta_4} + \dot{r}_{P2P3}e^{i\theta_3}$$

The velocity of P_2 in the XY system can be formulated separately from Figure 2.38(b) as

$$\dot{\mathbf{r}}_{P2} = i\dot{\theta}_2 r_{P2}e^{i\theta_2}$$

Combining the last two expressions yields

$$i\dot{\theta}_2 r_{P2}e^{i\theta_2} = \dot{r}_{P3}e^{i\theta_4} + \dot{r}_{P2P3}e^{i\theta_3}$$

which is the same expression obtained from the vector polygon for the geometry.

2.10 LOOP EQUATIONS

The closed polygon "loop" model used to formulate complex variable solutions in the last section can be applied to systems containing more than one loop. The system shown in Figure 2.39 is an example of such a system. There are six rigid links in this mechanism, since member (2) is a rigid link extending from 0_2 through A to B.

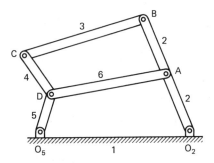

Figure 2.39   Six-Bar Mechanism

Two loop equations can be written for this mechanism using the polygons shown in Figures 2.40(a) and (b). The vector geometry equations for these loops are

$$\mathbf{r}_1 + \mathbf{r}_2 + \mathbf{r}_3 + \mathbf{r}_4 + \mathbf{r}_5 = \mathbf{0}$$

$$\mathbf{r}_2' + \mathbf{r}_3 + \mathbf{r}_4 + \mathbf{r}_6 = \mathbf{0}$$

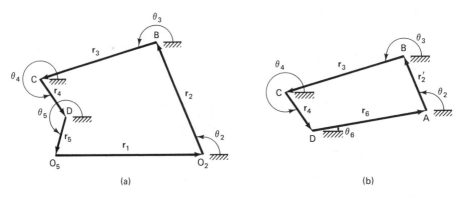

(a)   (b)

Figure 2.40   (a) Outer Loop, (b) Upper Loop

Subtracting these equations yields

$$\mathbf{r}_1 + \mathbf{r}_2 - \mathbf{r}_2' + \mathbf{r}_5 - \mathbf{r}_6 = \mathbf{0}$$

which is the equation for the redundant lower loop shown in Figure 2.41.

Each of the independent loop equations generates two equations, one for the horizontal projections and one for the vertical projections. For the sample mechanism, these would be

$$r_1 \cos \theta_1 + r_2 \cos \theta_2 + r_3 \cos \theta_3 + r_4 \cos \theta_4 + r_5 \cos \theta_5 = 0$$

$$r_1 \sin \theta_1 + r_2 \sin \theta_2 + r_3 \sin \theta_3 + r_4 \sin \theta_4 + r_5 \sin \theta_5 = 0$$

$$r_2' \cos \theta_2 + r_3 \cos \theta_3 + r_4 \cos \theta_4 + r_6 \cos \theta_6 = 0$$

$$r_2' \sin \theta_2 + r_3 \sin \theta_3 + r_4 \sin \theta_4 + r_6 \sin \theta_6 = 0$$

Since only the angles vary with time, the time derivatives of these equations will be linear in the angular velocities and angular accelerations. The coefficients of the velocities and accelerations are dependent on the known lengths and instantaneous angles of the links. There are five unknown angular velocities and accelerations ($\theta_1 = \dot{\theta}_1 = \ddot{\theta}_1 = 0$) and only four equations. Therefore, to be solvable, one angular veloc-

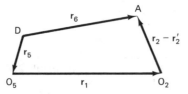

Figure 2.41   Lower Loop

ity and one angular acceleration must be given. Solutions will then exist, provided the determinant of the coefficients does not vanish.

If point A is allowed to slide along member (2) instead of being fixed, then the velocity and acceleration equations would also have terms containing $\dot{r}_{2''}$ and $\ddot{r}_{2'}$. Since the number of equations has not been increased, these velocities and accelerations can be found only if two velocities and accelerations are given.

There are occasionally certain system configurations for which a solution cannot be obtained even when a sufficient number of velocities and accelerations are prescribed. An example of such a mechanism is shown in Figure 2.42.

The velocity equations for this device are

$$-r_2 \dot{\theta}_2 \sin\theta_2 - r_3 \dot{\theta}_3 \sin\theta_3 = \dot{r}_4$$

$$r_2 \dot{\theta}_2 \cos\theta_2 + r_3 \dot{\theta}_3 \cos\theta_3 = 0$$

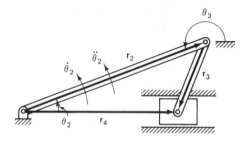

Figure 2.42 Slider Crank Mechanism

Assuming that $\dot{\theta}_2$ is given, then the equations for $\dot{\theta}_3$ and $\dot{r}_4$ can be written in the form: a kinematic matrix times a column vector of the unknowns equals a column vector of the inputs, i.e.,

$$\begin{bmatrix} r_3 \sin\theta_3 & 1 \\ -r_3 \cos\theta_3 & 0 \end{bmatrix} \begin{bmatrix} \dot{\theta}_3 \\ \dot{r}_4 \end{bmatrix} = r_2 \dot{\theta}_2 \begin{bmatrix} -\sin\theta_1 \\ \cos\theta_2 \end{bmatrix}$$

In this case, the kinematric matrix will be "singular" when its determinant is zero, which occurs when $\theta_3 = \pm\pi/2,\ 3\pi/2,\ \ldots$ When θ_3 assumes any of these angles, the mechanism "locks up" if $r_2 > r_3$. Matrix singularities warn of critical configurations that require special consideration. In this case, lockup could be accompanied by excessive stresses.

PROBLEMS

With the exception of the complex variable problems, it is intended that the following exercise problems be solved using the graphical methods developed in this chapter. To facilitate their solution it is suggested that the figures be photocopied and attached to the top of the solution page. All constructions can then be referenced to the geometry of the mechanism.

2.1. Starting with the general acceleration equation, obtain the acceleration of point P in cylindrical coordinates using the moving orthogonal-base vectors $\mathbf{e}_r$, $\mathbf{e}_\phi$, and $\mathbf{e}_z$ shown in the diagram. Scalar r is the projection of P on the XY plane. Vector $\mathbf{e}_r$ is fixed to r. Vector $\mathbf{e}_z$ is the same as $\mathbf{k}$ in rectangular coordinates.

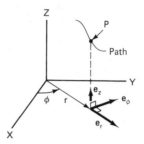

2.2. Bead P shown in the diagram moves radially inward along a spoke at a constant 5 m/s while the wheel moves forward at a constant angular velocity of $\omega = 10$ rad/s. Using graphical methods, find the acceleration of the bead as shown if it is 0.1 m from the center of the hub.

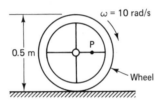

2.3. A particle P moves along the curve $Y = \frac{1}{2}X^2$, as shown in the diagram. The path length measured from the coordinate origin is $s = t^2$, where s is in feet, and t is in seconds. When $X = 1.0$ feet, find the following:
(a) The length s and the time.
(b) The vectors $\mathbf{e}_n$ and $\mathbf{e}_t$.
(c) The radius of curvature.
(d) The particle acceleration.

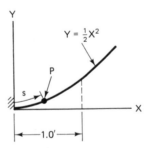

2.4. Link (2) in the four-bar linkage shown in the diagram has a constant angular velocity of 2.9 rad/s (CCW). Using graphical methods, find the angular velocity of link (4).

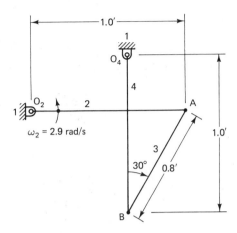

2.5. The four-bar linkage shown in the diagram has a known velocity of magnitude $V_B = 11.0$ cm/s. Using graphical methods, find $\mathbf{V}_A$, ω_2, ω_3, and ω_4.

$AO_4 = 1.5$ cm
$BO_2 = 1.75$ cm
$\;AB = 3.75$ cm

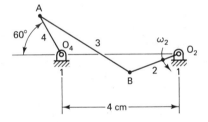

2.6. A slider–crank mechanism is shown in the diagram. The slider moves to the right with an instantaneous acceleration $A_s = 1.0$ ft/s^2 and a velocity of $V_s = 1.0$ ft/s. Find ω_2, $\dot{\omega}_2$, ω_3, and $\dot{\omega}_3$ at this instant.

$AO_2 = 2.24$ ft
$\;AB = 1.0$ ft
$BO_2 = 2.0$ ft

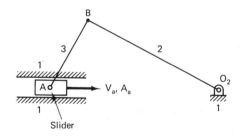

2.7. Consider the four-bar linkage shown in the diagram.

$O_2A = 1.2$ in. $\omega_2 = 1$ rad/s (CW)
$AB = 4.5$ in. $\dot{\omega}_2 = 2$ rad/s^2 (CCW)
$O_4B = 2.0$ in. $CB = 2.8$ in.
$DC = 2.0$ in. $AC = 2.3$ in.

(a) Determine ω_3, ω_4, and $\mathbf{V}_C$. Use a graphical solution. Define all quantities including the length and velocity scales.

(b) Find $\dot{\omega}_3$, $\dot{\omega}_4$, and $\mathbf{A}_C$.

(c) Find the velocity and acceleration of the slider at point D.

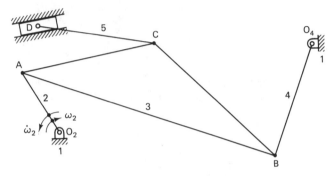

2.8. The disk (3) shown in the diagram rotates at a constant angular velocity $\omega_3 = 2.0$ rad/s (CW). A slider in a slot in the disk causes rod (2) to rotate. When the slider (S) is 1.0 ft from O_3, find relative velocity and acceleration of the slider with respect to the slot and ω_2 and $\dot{\omega}_2$. Use graphical methods.

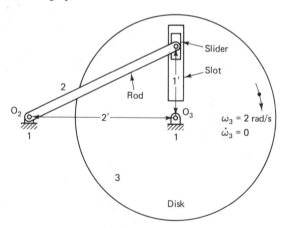

2.9. Using graphical methods, determine $\dot{\omega}_4$ and $\dot{\omega}_5$.

$O_2A = \frac{1}{2}$ in. $BC = 2$ in.
$AB = 2$ in. $O_5C = \frac{1}{2}$ in.

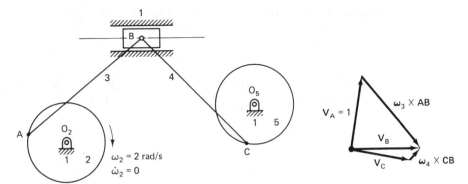

2.10. The diagram shows a slider–crank mechanism in which slider (2) drives crank (3); slider (2) has a curved slot cut in it with a radius of curvature of 5 cm. Using graphical methods, find ω_3 and $\dot{\omega}_3$.

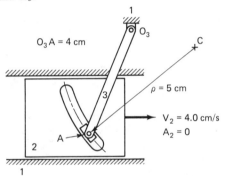

2.11. Disk (2) rolls to the right such that the rightward velocity of point A is a constant 1 in./s. Distance BA is 0.6 in. Point B does not slip (point C does slip). Determine the magnitude and direction of the following:

(a) ω_3

(b) V_{D5}

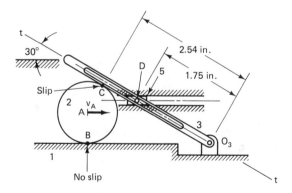

2.12. Consider a slider mechanism shown in the figure drawn to scale.

$O_2A = 2.0$ in.

$O_3A = 0.5$ in.

$R_C = 0.75$ in.

$\omega_2 = 5.0$ rad/s

(a) Find the magnitude and direction of the angular velocity of the output link (3).

(b) Find the magnitude and direction of the relative velocity of point A_2 with respect to A_3.

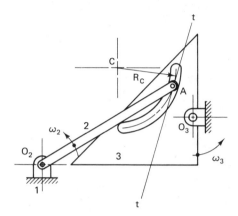

2.13. The slider shown in the diagram moves with a constant velocity of 1.0 in./s. Link (4) has a constant angular velocity. The relative velocity of the pin at B_4 along the curved slot in plate (3) is zero. Determine the magnitude and direction of ω_3 and $\dot{\omega}_3$ using graphical methods. Point C_3 is the center of the radius of curvature of the slot.

$\omega_4 = $ constant $BO_3 = 1.0$ in.

$AB = 2.0$ in. $C_3B_3 = 0.75$ in.

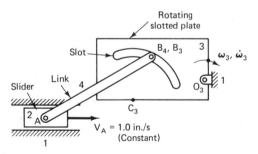

Drawing is to scale

2.14. Find the velocity and acceleration of the slider in the diagram when $\rho = 2.3$ cm, $v_{\text{rod}} = 1.0$ cm/s, $a_{\text{rod}} = 2.0$ cm/s^2, $\omega_2 = 1.0$ rad/s (constant), and $O_2P = 4$ cm. The rod rotates with $\boldsymbol{\omega}_2$ and slides through the bushing with velocity $\mathbf{v}_{\text{rod}}$ and acceleration $\mathbf{a}_{\text{rod}}$ relative to the bushing.

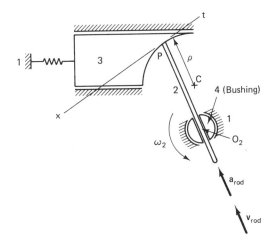

2.15. The diagram shows two slotted links, (2) and (4), joined by a pin (3). The radius of curvature of the curved slot is ρ_4. The velocity diagram is shown below it. Using graphical methods, obtain the acceleration of the center of pin P_3.

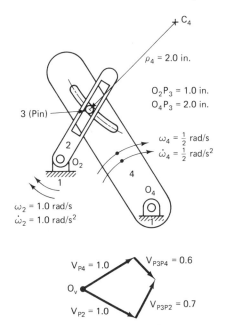

2.16. For the mechanism shown in the diagram and the conditions given, find the following:
  (a) Let $R_A = \infty$, $V_{B3} = 0$, and $A'_{B3} = 0$. Find ω_3, $\dot{\omega}_3$, V_{A3}, and A_{A3}.
  (b) Let $R_A = 1$ cm, $V_{B3} = 20$ cm/s, and $A'_{B3} = 30$ cm/s². Repeat part (a).

For parts (a) and (b):

$O_2A = 5.5$ cm $\omega_2 = 2$ rad/s (CCW)

$AB = 4.8$ cm $\dot{\omega}_2 = 3$ rad/s² (CCW)

$R_B = 3.7$ cm

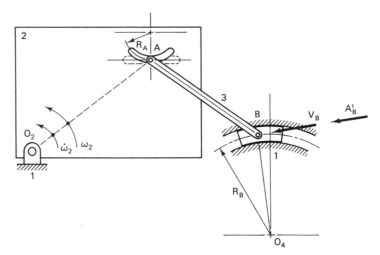

2.17. A pin P slides at the point of common intersection shown in the diagram.

    (a) Construct the velocity diagram to scale.

    (b) Find the acceleration of pin 4 using graphical methods. Pin 4 is free to slide simultaneously in plates 2 and 3.

    For parts (a) and (b):

    $\omega_2 = 1.0$ rad/s (constant) $\omega_3 = 0.5$ rad/s (constant)

    $O_2P = 1.6$ in. $R_2 = 1.0$ in.

    $O_3P = 1.5$ in. $R_3 = 0.7$ in.

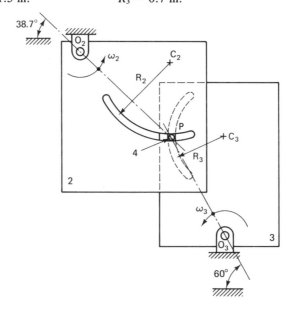

2.18. For the mechanism shown in the diagram, find $\boldsymbol{\omega}_3$ and $\dot{\boldsymbol{\omega}}_3$ for

    (a) the straight slot configuration

    (b) the curved slot configuration

    $O_4B = 1$ cm

     $AB = 1.5$ cm

      $\omega_4 = 4$ rad/s constant

      $\omega_2 = 10$ rad/s

      $\dot{\omega}_2 = 30$ rad/s²

    Scale 1 cm = 0.2 cm

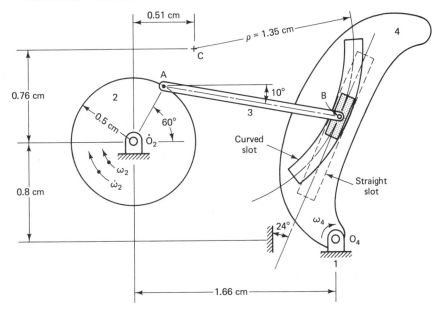

2.19. The diagram shows a cam (2) moving to the right with a constant velocity $V_2 = 2.0$ in./s. Disk (3) rolls without slipping on the circular face of the cam, causing the follower (4) to move vertically. Using graphical methods, find the angular velocity and acceleration of the disk, and the velocity and acceleration of the follower (4).

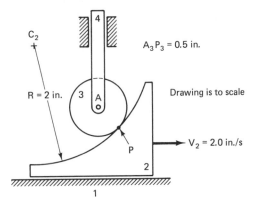

2.20. A cylinder (2) rolls without slipping on the curved stationary surface at P with the angular velocity and angular acceleration indicated in the diagram. The radius of curvature of the stationary surface is 2.5 in. The slotted link (3) is pinned to the center of the disk at A and slides on pin B, which is fixed to ground. The compressed spring k causes the rolling motion. Using graphical methods, find $\boldsymbol{\omega}_3$ and $\dot{\boldsymbol{\omega}}_3$.

$\omega_2 = 10.0$ rad/s $\quad PC = 2.5$ in.

$\dot{\omega}_2 = 5.0$ rad/s $\quad AB = 2.0$ in.

$AP = 0.5$ in.

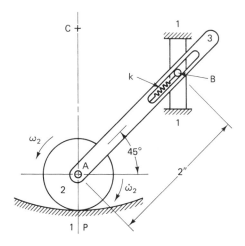

2.21. Using graphical methods, obtain $\dot{\boldsymbol{\omega}}_3$ when $\omega_2 = 15$ rad/s (constant). See the diagram.

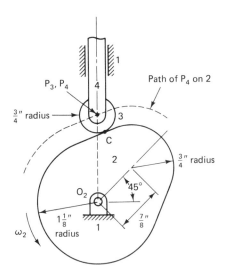

2.22. The translating cam-follower system and its velocity diagram are shown in the figure. Using graphical methods, find the angular acceleration of the roller (3), i.e., both its magnitude and sense.

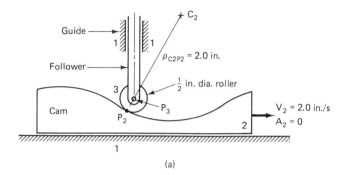

(a)

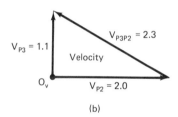

(b)

2.23. The diagram shows a planetary, or epicyclic, gear train. The angular velocity and acceleration of the carrier (2) are $\omega_2 = 12$ rad/s and $\dot{\omega}_2 = 48$ rad/s^2, respectively. Using graphical methods, determine $\boldsymbol{\omega}_3$ and $\dot{\boldsymbol{\omega}}_3$ of the planetary gear (3). The outer gear (1) is fixed to ground.

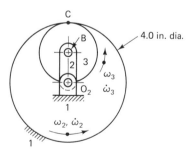

2.24. Rollers (4) and (5) contact each other without slipping at C in the diagram. Using graphical methods, find $\boldsymbol{\omega}_4$ and $\dot{\boldsymbol{\omega}}_4$ at the instant shown.

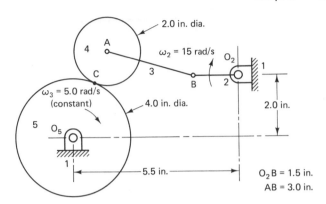

2.25. The diagram, which is drawn to scale, shows two cams whose surfaces slide relative to each other at their contact point P. Cam (2) is driven at $\omega_2 = 1.0$ rad/s. Using graphical methods, determine ω_3 and the relative velocity of the contacting surfaces, i.e., V_{P2P3}.

$O_2 P_2 = 7$ cm $O_2 C_2 = 5$ cm
$O_3 P_3 = 6.5$ cm $O_3 C_3 = 5$ cm
$\omega_2 = 1$ rad/s

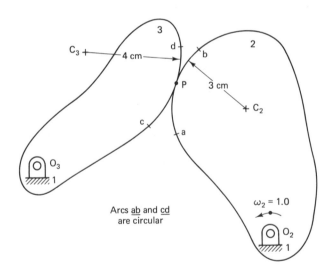

2.26. Using the equivalent-mechanism method find ω_4, $\dot{\omega}_4$, ω_3, $\dot{\omega}_3$, V_{P3}, and A_{P3} for the mechanism in the diagram.

$\omega_2 = 0.5$ rad/s = constant $R = 2$ cm $= AP_3$
$AO_4 = 6$ cm $O_2 P = 5.5$ cm

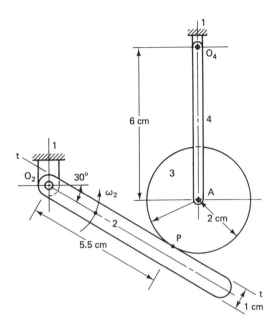

2.27. The diagram shows a cam follower (2) that contacts the cam (3) at point A. Contact is maintained by a coil spring attached to (3), which is not shown. Devise an equivalent mechanism for this device. Using graphical methods and the equivalent mechanism, find ω_3, $\dot{\omega}_3$, V_{A2A3}, and A_{A2A3}.

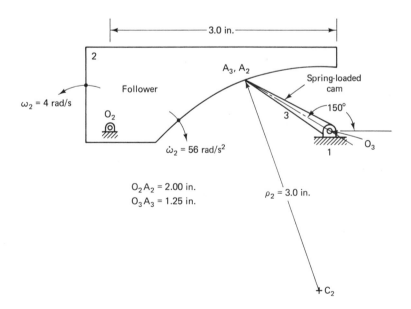

2.28. The diagram shows a roller-follower mechanism. No slip occurs at the contact points P_3 and P_4. The arm (2) rotates at a constant angular velocity. The follower has an upward velocity of 1.1 in./s and a downward acceleration of 2.0 in./s². Using graphical methods, obtain the angular acceleration of the roller (3). State size and direction. Point C_4 is the center of the radius of curvature of the follower face.
$A_3 P_3 = 0.5$ in.
$C_4 P_4 = 0.5$ in.
$O_2 A_2 = 1.0$ in.

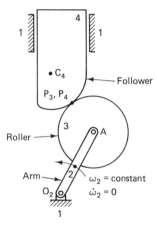

2.29. Solve Problem 2.23 using complex-variable methods.

2.30. Solve Problem 2.17 using complex-variable methods.

2.31. Bar $O_2 B$ is rotating clockwise at a constant angular velocity of 5 rad/s; see the diagram. Bar BGD slides through frictionless bearings at E and F. Using the method of complex variables, find the angular velocity of member EFO_4.

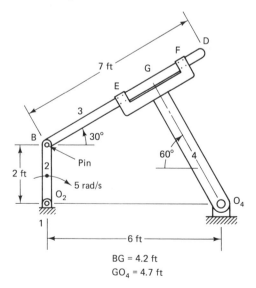

BG = 4.2 ft
GO$_4$ = 4.7 ft

2.32. A cam-follower mechanism is illustrated in the diagram. The velocity and the acceleration of the follower are shown. The cam is an eccentrically mounted circular disk whose center (C_2) is level with its pivot (O_2) at this instant. Using complex-variable methods, find ω_2 and $\dot\omega_2$.

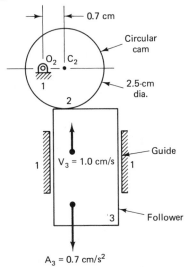

2.33. The disk shown in the diagram rotates with $\omega_2 = 1$ rad/s (CW). A circular slot with radius of 2 in. is cut into the disk. The center of curvature is at C_2, 1 in. from the disk center. A block, attached to the end of link (3), slides in the curved slot, causing link (3) to rotate. Construct an equivalent mechanism for this device and determine the angular velocities of any links found in the equivalent mechanism that are not found in the original device. Use complex-variable methods to perform your analysis.

$\rho = C_2 A_2 = 2$ in.
$O_2 C_2 = 1$ in.
$O_3 A_3 = 3$ in.

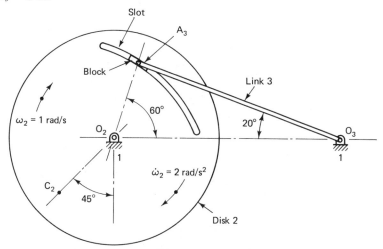

2.34. A link (3) drives a disk (2) through a slider B in a slot in disk (2) as shown in the diagram. Using complex-variable methods, determine the velocity of slider B relative to the slot.

$O_3 B = 3$ in.
$O_2 A_2 = 1$ in.
$A_2 B = 1$ in.

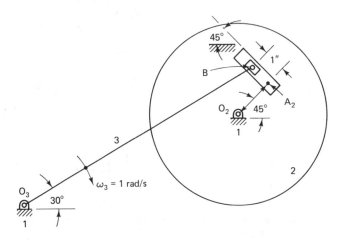

2.35. In the diagram, disk (3) rolls with $\omega_3 = 1.0$ rad/s without slipping at contact point C. Rod AB slides through a pivoting tube. The tube is pinned to ground at A. Using complex-variable methods, determine the rate at which the distance between the pivot at A and the center of the disk at B changes.

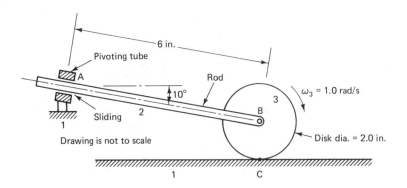

2.36. The diagram shows a cam-roller-follower system. Using complex variables, determine the vertical velocity of the follower (4) at the instant shown.

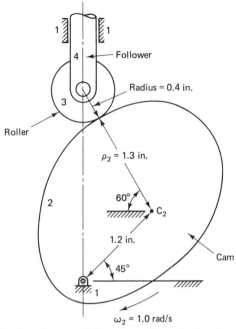

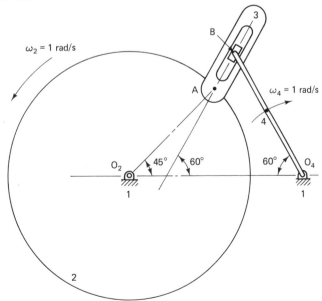

2.37. The slotted arm (3) is pinned to the disk at A, as shown in the diagram. Slider B at the end of the rigid link (4) slides in the slot. Using complex-variable methods, set up the scalar equations for the angular velocities of members (3) and (4). Insert the known inputs into these equations, but do not solve them.

$O_2A = 1$ in.
$AB = \frac{1}{2}$ in.
$O_4B = 1.32$ in.

2.38. With the proper adjustment of link lengths, the drag-link device shown in the diagram can be used as a quick-return mechanism. Write the loop equations for the system and obtain the equations for the velocities and accelerations. Determine how many kinematic inputs are required. Provide a sketch defining all the symbols used.

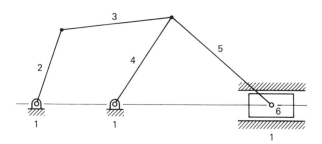

2.39. Assuming that the velocity of slider 6 in Problem 2.38 is not given (i.e., it is not a driver, but driven), find a configuration for the mechanism that is singular for velocity. Do this by inspection of the diagram of Problem 2.38. Verify that the velocity equations are mathematically singular for this configuration.

2.40. The mechanism in the diagram has an obvious "singular configuration" where it will lock up. Using complex-variable techniques, show that $\dot{\theta}_3$ becomes mathematically infinite in this configuration.

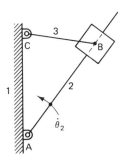

2.41. The mechanism shown in the diagram is used to drive the slider block (3). Although numbered separately, (1) and (4) are equal-length portions of a solid link 2.0 in. long. This link is given a constant counterclockwise angular velocity of 1.0 rad/s, causing slider (3) to move horizontally. Link (6) is vertical at this instant. Using complex-variable methods, determine the following:
(a) the lengths of links (5) and (6)
(b) the angular velocity of link (2)

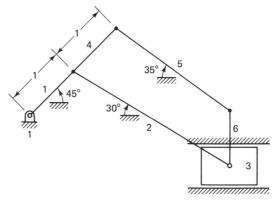

Length of link (2) = 3

REFERENCES

1. Shames, I. H. *Engineering Mechanics—Statics and Dynamics,* 3rd Ed. Englewood Cliffs, NJ: Prentice Hall, 1980.
2. Mabie, H. H., and Ocvirk, F. W. *Mechanisms and Dynamics of Machinery,* 3rd Ed. New York: John Wiley, 1975.

3

DYNAMICS

3.1 INTRODUCTION

There are many machine dynamic systems where the kinematics of the elements are established before the forces and couples required to produce the motions are known. The examples of the last chapter are in this category.

When this is the case, the designer may proceed directly to the dynamic analysis of the system components to determine the loads on the various parts. Then, by employing elastic theory or strength-of-materials concepts, the stresses and deflections of the parts can be predicted. If these are not satisfactory, adjustments in the design of the part can be made and the process repeated until the design criteria have been met.

In this chapter, we will develop a graphical method of dynamic analysis applicable to planar mechanisms. This method complements the graphical kinematic technique of the previous chapter, providing a quick method for finding interconnecting loads and, in certain cases, internal loads and deflections.

3.2 NEWTON'S SECOND LAW FOR A SYSTEM OF PARTICLES

For a single particle of constant mass m_i subject to external forces $\mathbf{F}_i$ and internal forces $\mathbf{f}_{ij}$, Newton's second law is

$$\sum_{j=1}^{N} \mathbf{f}_{ij} + \mathbf{F}_i = \frac{d}{dt}(m_i \dot{\mathbf{r}}_{oi}) \tag{3.1}$$

where $\dot{\mathbf{r}}_{oi}$ is the position vector of m_i measured relative to an inertial reference frame; see Figure 3.1. The summation $\sum_{j=1}^{N} \mathbf{f}_{ij}$ is the resultant force exerted on m_i by all of the other particles that are internal to the system of particles. In a solid, this vector sum would represent the net effect of all the other molecules on the ith molecule.

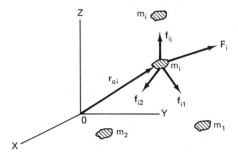

Figure 3.1   Forces on Mass m_i

The vector $\mathbf{F}_i$ is the net effect of all the forces external to the system of particles. These forces fall into two general categories: those due to contact with another system of particles, e.g., between connecting parts of a machine; and those due to action at a distance, such as the earth's gravitational force.

When dealing with a large collections of particles, it is neither possible nor practical to consider examining the motion of each individual particle using the form of Newton's law just given. If the equation is extended to the entire collection of particles comprising the system by summing over all the particles in the collection, Newton's second law takes the form

$$\sum_{i=1}^{N} \sum_{j=1}^{N} \mathbf{f}_{ij} + \sum_{i=1}^{N} \mathbf{F}_i = \sum_{i=1}^{N} \frac{d}{dt}(m_i \dot{\mathbf{r}}_{oi})$$

In the double summation, the force of the ith particle upon itself is zero, i.e., $\mathbf{f}_{ii} = 0$. When the forces between particles are equal and opposite, then these forces occur in self-cancelling pairs so that $\mathbf{f}_{ij} = -\mathbf{f}_{ji}$. As a result, the double sum vanishes for the system of particles, simplifying the left-hand side of that equation so that it reduces to

$$\sum_{i=1}^{N} \mathbf{F}_i = \sum_{i=1}^{N} \frac{d}{dt}(m_i \dot{\mathbf{r}}_{oi})$$

It should be noted that this form of Newton's law results from an assumption about the internal forces. The remaining term on the left-hand side of the equation is the sum of all the external forces acting on the system. The introduction of additional information regarding the nature of the internal forces has reduced the left side of the original equation to a form that is directly applicable to practical problems.

Unfortunately, the same is not true of the right-hand side, which still calls for the examination of the motion of each individual particle, m_i. The right-hand side can be rearranged to take the form

$$\sum_{i=1}^{N} \frac{d}{dt}(m_i \dot{\mathbf{r}}_{oi}) = \frac{d^2}{dt^2} \sum_{i=1}^{N} m_i \mathbf{r}_{oi}$$

if the mass of each particle is assumed to remain constant. Although this manipulation may seem trivial, it leads to another simplifying concept called the center of mass. The sum represented by $\sum_{i=1}^{N} m_i \mathbf{r}_{oi}$ is a purely geometric quantity representing the mass-weighted sum of the position vectors of all the particles m_i at a given instant of time. The center of mass is defined as

$$\mathbf{R}_g = \frac{\sum_{i=1}^{N} m_i \mathbf{r}_{oi}}{\sum_{i=1}^{N} m_i} = \frac{\sum_{i=1}^{N} m_i \mathbf{r}_{oi}}{M} \tag{3.2}$$

where M is the total mass of all the particles.

The symbol $\mathbf{R}_g$ was chosen here because we will find it convenient in the development of the moment-of-momentum equation to place the origin of an auxiliary xyz coordinate system at g. Vector $\mathbf{R}_g$ will then also be the position vector for the origin of that coordinate system.

For rigid solids, the center of mass g is a point fixed in the solid. For real solids that are nearly rigid, g moves very little under loads that do not permanently deform the material.

The center-of-mass concept allows the right-hand side of the equation to be written as

$$\frac{d^2}{dt^2} \sum_{i=1}^{N} m_i \mathbf{r}_{oi} = M\ddot{\mathbf{R}}_g$$

With the aid of the center-of-mass concept, Newton's law for a rigid body takes the form

$$\sum_{i=1}^{N} \mathbf{F}_i = M\ddot{\mathbf{R}}_g \tag{3.3}$$

which states that the sum of the external forces equals the mass of the system times the acceleration of the center of mass.

This restricted version of Newton's second law is well suited to the study of the motion of rigid and quasirigid bodies. Bear in mind, however, that it is applicable only to situations where the mass is constant and the internal forces are self-canceling.

3.3 MOMENT OF MOMENTUM FOR A SYSTEM OF PARTICLES

In its final version, Newton's law for a rigid body focuses attention on a single point and the effect of the external forces on its motion. From our study of kinematics, we know that a complete description of rigid-body motion requires a description of the motion of some fixed *point* in the body *and* the body's *angular velocity* and *acceleration*. Newton's law provides a means of determining the kinematics of a particular point, the center of mass, but reveals nothing about the angular velocity or acceleration. Actually, information about the motion of other points in the body was sacrificed by the introduction of the center-of-mass concept.

From our study of statics, we know that equilibrium of a "free body" requires that the vector sum of the external forces *and* moments must both be zero. Newton's law for a rigid body is obviously the counterpart of the static-force equilibrium equation, where the dynamic effect is embodied in the center-of-mass acceleration times the mass term of the right-hand side. By analogy, we would also expect there to be a dynamic counterpart to the static-moment equilibrium equation. This analogy suggests where we might look for the missing equation, that is, in the formulation of a dynamic-moment equation.

At first sight, the creation of a dynamic-moment equation would seem to be impossible since there is only one dynamic equation for a particle, Newton's second law, and we have already used that law in the previous section. The question then is, "how do we create an independent moment formulation starting again with Newton's second law?" The answer is simple and at the same time as profound as any in all of physics. It is to "introduce additional information into the formulation." It is obvious that if that is not done, we will surely only succeed in rederiving the rigid-body equation we already have.

Every branch of engineering and physics is replete with examples of the expansion of the fundamental laws via the introduction of additional information. The question to be answered in each case is, what information? In the case of rigid-body dynamics, it will actually be a further restriction on an assumption that has already been made.

Figure 3.2 shows a rigid body with points g and a fixed in the body. The moving coordinates xyz are also fixed in the body so that the body and the xyz coordinates move together with $\mathbf{\Omega} = \mathbf{\omega}$ and $\dot{\mathbf{\Omega}} = \dot{\mathbf{\omega}}$. Typical particle m_i is also fixed in the solid and located relative to xyz by $\mathbf{r}_i$. External and internal forces $\mathbf{F}_i$ and $\mathbf{f}_{ij}$, respectively, act on mass m_i.

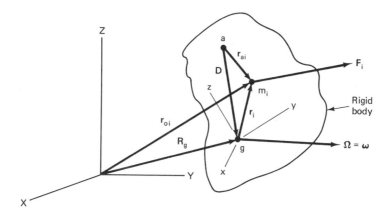

Figure 3.2   Rigid Body with Moving Coordinates Fixed in the Body at g

The starting point for the developement is once again the only fundamental law we have for a particle, Newton's second law. The moment of the internal and external forces in this equation will be taken about arbitrary point a in Figure 3.2. To accomplish this, Equation (3.1) will be crossed with the $\mathbf{r}_{ai}$ and summed over all particles in the system. Then

$$\sum_{i=1}^{N} \mathbf{r}_{ai} \times \mathbf{F}_i + \sum_{i=1}^{N} \sum_{j=1}^{N} \mathbf{r}_{ai} \times \mathbf{f}_{ij} = \sum_{i=1}^{N} \mathbf{r}_{ai} \times \frac{d}{dt}(m_i \dot{\mathbf{r}}_{oi})$$

Consistent with our previous assumption, the internal forces will be taken to be self-cancelling, in which case the double summation will contain paired internal moments of the type

$$\mathbf{r}_{ai} \times \mathbf{f}_{ij} + \mathbf{r}_{aj}\mathbf{f}_{ji} = (\mathbf{r}_{ai} - \mathbf{r}_{aj}) \times \mathbf{f}_{ij}$$

The vector difference $\mathbf{r}_{ai} - \mathbf{r}_{aj}$ is directed along the line joining the two particles.

We will now make an additional assumption that adds information to the analysis. It will be assumed that the internal forces are not only equal and opposite, but that they lie along the line joining the particles. If the internal forces $\mathbf{f}_{ij}$ act along a line joining masses m_i and m_j, then the moment $(\mathbf{r}_{ai} - \mathbf{r}_{aj}) \times \mathbf{f}_{ij} = \mathbf{0}$, since the vectors are parallel.

Previously, it was assumed that the internal forces between two masses were equal and opposite, but it was not necessary to know the line action of either of the forces. The additional assumption introduced here is that they lie along a particular direction. Fortunately, a mass in the gravity field created by other masses does experience mutually attractive forces that behave this way. Although the nature of the internal forces has been further restricted by the additional assumptions, the resulting model still conforms to a wide variety of real materials.

From Figure 3.2, the moment-arm vector $\mathbf{r}_{ai}$ can be written as

$$\mathbf{r}_{ai} = \mathbf{D} + \mathbf{r}_i$$

where $\mathbf{D}$ locates the center of mass relative to point a. The velocity of mass m_i can also be written as

$$\dot{\mathbf{r}}_{oi} = \dot{\mathbf{R}}_g + \dot{\mathbf{r}}_i$$

By using these, the right-hand side of the moment equation can be rearranged as follows:

$$\sum_{t=1}^{N} \mathbf{r}_{ai} \times \frac{d}{dt}(m_i\dot{\mathbf{r}}_{oi}) = \sum_{i=1}^{N} (\mathbf{D} + \mathbf{r}_i) \times \frac{d}{dt}(m_i\dot{\mathbf{r}}_{oi})$$

$$= \sum_{i=1}^{N} \mathbf{D} \times \frac{d}{dt}(m_i\dot{\mathbf{r}}_{oi}) + \sum_{i=1}^{N} \mathbf{r}_i \times \frac{d}{dt}(m_i\dot{\mathbf{R}}_g) \qquad (3.4a)$$

$$+ \sum_{i=1}^{N} \mathbf{r}_i \times \frac{d}{dt}(m_i\dot{\mathbf{r}}_i)$$

In the first term, the center-of-mass concept allows the derivative to be replaced with

$$\mathbf{D} \times \sum_{i=1}^{N} \frac{d}{dt}(m_i\dot{\mathbf{r}}_{oi}) = \mathbf{D} \times M\frac{d}{dt}(\dot{\mathbf{R}}_g) = \mathbf{D} \times M\ddot{\mathbf{R}}_g$$

Carrying out the indicated operations of the second term yields

$$\sum_{i=1}^{N} \mathbf{r}_i \times \frac{d}{dt}(m_i\dot{\mathbf{R}}_g) = \left(\sum_{i=1}^{N} \mathbf{r}_i m_i\right) \times \frac{d}{dt}(\dot{\mathbf{R}}_g)$$

The sum $\Sigma\, \mathbf{r}_i m_i = \mathbf{0}$ since $\mathbf{r}_i$ emanates from the center of mass.

Before examining the last term, it will be noted that the product rule of differentiation yields

$$\frac{d}{dt}\sum_{i=1}^{N} (\mathbf{r}_i \times m_i\dot{\mathbf{r}}_i) = \sum_{i=1}^{N} (\dot{\mathbf{r}}_i \times m_i\dot{\mathbf{r}}_i) + \sum_{i=1}^{N} \left[\mathbf{r}_i \times \frac{d}{dt}(m_i\dot{\mathbf{r}}_i)\right]$$

The quantity $\dot{\mathbf{r}}_i \times m_i\dot{\mathbf{r}}_i = \mathbf{0}$ since it is the cross product of the same vector. Now the last term can be written

$$\sum_{i=1}^{N} \mathbf{r}_i \times \frac{d}{dt}(m_i\dot{\mathbf{r}}_i) = \frac{d}{dt}\sum_{i=1}^{N} (\mathbf{r}_i \times m_i\dot{\mathbf{r}}_i)$$

With these substitutions, the entire right-hand side of the moment equation reduces to

$$\sum_{i=1}^{N} \mathbf{r}_{ai} \times \frac{d}{dt}(m_i \dot{\mathbf{r}}_{oi}) = \mathbf{D} \times M\ddot{\mathbf{R}}_g + \frac{d}{dt} \sum_{i=1}^{N} (\mathbf{r}_i \times m_i \dot{\mathbf{r}}_i) \qquad (3.4b)$$

The first term, $\mathbf{D} \times M\ddot{\mathbf{R}}_g$, is just the moment of the external forces about a, since $\sum_{i=1}^{N} \mathbf{F}_i = M\ddot{\mathbf{R}}_g$ by Newton's second law for a rigid body.

The second term is a moment of momentum of a very special kind. This moment is about the center of mass and is part of the total momentum for a particle, which is

$$m_i \dot{\mathbf{r}}_{oi} = m_i \dot{\mathbf{R}}_g + m_i \dot{\mathbf{r}}_i$$

or, canceling the particle mass,

$$\dot{\mathbf{r}}_{oi} = \dot{\mathbf{R}}_g + \dot{\mathbf{r}}_i$$

From our study of kinematics, we know that the particle velocity can also be written as

$$\dot{\mathbf{r}}_{oi} = \dot{\mathbf{R}}_g + \boldsymbol{\Omega} \times \mathbf{r}_i + \mathbf{v}_{xyz}$$

If the moving coordinate system is fixed in the rigid body then $\mathbf{v}_{xyz} = \mathbf{0}$. Comparing the two expressions shows that when $\boldsymbol{\Omega} = \boldsymbol{\omega}$

$$\dot{\mathbf{r}}_i = \boldsymbol{\omega} \times \mathbf{r}_i$$

We are now in a position to write the final form of the dynamic-moment equation for a system of discrete masses. Defining the "moment-of-momentum" vector as

$$\mathbf{H}_g = \sum_{i=1}^{N} m_i \mathbf{r}_i \times (\boldsymbol{\omega} \times \mathbf{r}_i)$$

then

$$\sum_{i=1}^{N} \mathbf{r}_{ai} \times \mathbf{F}_i = \mathbf{D} \times M\ddot{\mathbf{R}}_g + \dot{\mathbf{H}}_g \qquad (3.5)$$

In words, this equation states that "the moment of the external forces about an arbitrary point a equals the moment about a of the sum of the external forces (acting at g) plus the time rate of change of a 'moment of momentum' about the center of mass."

An important feature of the right-hand side of this equation is that everything is referred to the center of mass g. This feature becomes especially important when it comes to evaluating the moment-of-momentum term since it "standardizes" the way this term is formulated.

To illustrate this, consider the system to be a continuum rather than a collection of discrete masses. Then

$$\sum \longrightarrow \iiint$$

and

$$m_i \longrightarrow dm$$

so that the moment-of-momentum vector relative to the center of mass is

$$\mathbf{H}_g = \sum_{i=1}^{N} m_i \mathbf{r}_i \times (\boldsymbol{\omega}_i \times \mathbf{r}_i) \longrightarrow \iiint_M \mathbf{r}_i \times (\boldsymbol{\omega}_i \times \mathbf{r}_i) \, dm$$

where

$$\mathbf{r}_i = x\mathbf{i} + y\mathbf{j} + z\mathbf{k}$$

$$\boldsymbol{\omega} = \omega_x\mathbf{i} + \omega_y\mathbf{j} + \omega_z\mathbf{k}$$

The expanded form of this integral is

$$\begin{aligned}
\mathbf{H}_g &= \iiint_M \mathbf{r}_i \times (\boldsymbol{\omega} \times \mathbf{r}_i) \, dm \\
&= \mathbf{i}\left[\omega_x \iiint_M (y^2 + z^2) \, dm - \omega_y \iiint_M xy \, dm - \omega_z \iiint_M zx \, dm \right] \\
&+ \mathbf{j}\left[-\omega_x \iiint_M xy \, dm + \omega_y \iiint_M (z^2 + x^2) \, dm - \omega_z \iiint_M yz \, dm \right] \\
&+ \mathbf{k}\left[-\omega_x \iiint_M zx \, dm - \omega_y \iiint_M yz \, dm + \omega_z \iiint_M (x^2 + y^2) \, dm \right]
\end{aligned} \tag{3.6}$$

Observe that each of the terms in the expanded version consists of the product of a kinematic quantity, ω_x, ω_y, or ω_z, and a geometric quantity, a moment or product of inertia. The moments and products of inertia are usually abbreviated as follows

$$I_{xx} = \iiint_M (y^2 + z^2) \, dm \qquad I_{xy} = I_{yx} = \iiint_M xy \, dm$$

$$I_{yy} = \iiint_M (x^2 + z^2) \, dm \qquad I_{xz} = I_{zx} = \iiint_M zx \, dm \tag{3.7}$$

$$I_{zz} = \iiint_M (x^2 + y^2) \, dm \qquad I_{yz} = I_{zy} = \iiint_M yz \, dm$$

The integrals are evaluated using coordinates fixed in the body with their origin at the center of mass. They are properties of the body that can be evaluated independently of the motion since they are independent of time. The effect of the mass distribution can, therefore, be expressed in a standardized form.

The kinematic time-dependent parts of these equations are the angular velocity components ω_x, ω_y, and ω_z, and the unit vectors $\mathbf{i}$, $\mathbf{j}$, and $\mathbf{k}$, that move with the body.

We are now able to complete the evaluation of the moment-of-momentum term, which requires taking the time derivative. Noting that

$$\frac{d\dot\omega_x}{dt} = \ddot\omega_x \qquad \frac{d\omega_y}{dt} = \dot\omega_y \qquad \frac{d\omega_z}{dt} = \dot\omega_z$$

and

$$\frac{d\mathbf{i}}{dt} = \boldsymbol{\omega} \times \mathbf{i} \qquad \frac{d\mathbf{j}}{dt} = \boldsymbol{\omega} \times \mathbf{j} \qquad \frac{d\mathbf{k}}{dt} = \boldsymbol{\omega} \times \mathbf{k}$$

The last right-hand term now becomes

$$\begin{aligned}
\dot{\mathbf{H}}_g &= \frac{d}{dt} \iiint_M \mathbf{r}_i \times (\boldsymbol{\omega} \times \mathbf{r}_i)\, dm \\
&= \mathbf{i}[\dot\omega_x I_{xx} + \omega_y\omega_z(I_{zz} - I_{yy}) + I_{xy}(\omega_z\omega_x - \dot\omega_y) \\
&\quad - I_{xz}(\dot\omega_z + \omega_y\omega_x) - I_{yz}(\omega_y^2 - \omega_z^2) \\
&\quad + \mathbf{j}[\dot\omega_y I_{yy} + \omega_z\omega_x(I_{xx} - I_{zz}) + I_{yz}(\omega_x\omega_y - \dot\omega_z) \\
&\quad - I_{yx}(\dot\omega_x + \omega_z\omega_y) - I_{zx}(\omega_z^2 - \omega_x^2)] \\
&\quad + \mathbf{k}[\dot\omega_z I_{zz} + \omega_x\omega_y(I_{yy} - I_{xx}) + I_{zx}(\omega_y\omega_z - \dot\omega_x) \\
&\quad - I_{zy}(\dot\omega_y + \omega_x\omega_z) - I_{xy}(\omega_x^2 - \omega_y^2)]
\end{aligned} \qquad (3.8)$$

This is the hoped-for result, i.e., an expression containing the angular-velocity and angular-acceleration components.

The continuum form of the time derivative of the moment of momentum is exceedingly complicated. Fortunately, it is rarely necessary to retain all the terms. If the xyz system is aligned with the principal coordinates so that $I_{xy} = I_{yz} = I_{xz} = 0$, then

$$\begin{aligned}
\dot{\mathbf{H}}_g &= \frac{d}{dt} \iiint_M \mathbf{r}_i \times (\boldsymbol{\omega} \times \mathbf{r}_i)\, dm \\
&= \mathbf{i}[I_{xx}\dot\omega_x - \omega_y\omega_z(I_{yy} - I_{zz})] \\
&\quad + \mathbf{j}[I_{yy}\dot\omega_y - \omega_z\omega_x(I_{zz} - I_{xx})] \\
&\quad + \mathbf{k}[I_{zz}\dot\omega_z - \omega_x\omega_y(I_{xx} - I_{yy})]
\end{aligned} \qquad (3.9)$$

For planar motion, where $\omega_x = \omega_y = \dot\omega_x = \dot\omega_y = 0$, then the result is the familiar term

$$\dot{\mathbf{H}}_g = \frac{d}{dt} \iiint_M \mathbf{r}_i \times (\boldsymbol{\omega} \times \mathbf{r}_i)\, dm = \mathbf{k} I_{zz}\dot\omega_z$$

3.4 D'ALEMBERT EQUATIONS OF MOTION FOR A RIGID BODY

D'Alembert rearranged Newton's second law and the dynamic-moment equation so that they are reminiscent of the static-equilibrium equations. D'Alembert's version of these equations for a rigid body are

$$\sum_{i=1}^{N} \mathbf{F}_i - M\ddot{\mathbf{R}}_g = 0 \tag{3.10}$$

$$\sum \mathbf{r}_{ai} \times \mathbf{F}_i - \mathbf{D} \times M\ddot{\mathbf{R}}_g - \dot{\mathbf{H}}_g = 0 \tag{3.11}$$

This arrangement treats the dynamic quantities $M\ddot{\mathbf{R}}_g$, $\mathbf{D} \times M\ddot{\mathbf{R}}_g$, and $\dot{\mathbf{H}}_g$ as a force, moment, and couple, respectively, applied to the body contrary to their positive sense. Summing to zero means that the body can be treated as a free body. More importantly, it means that the free-body techniques of statics can continue to be used if the static equations are amended to include the dynamic (inertial) terms shown in the d'Alembert equations.

To illustrate the use of the d'Alembert equations, consider the cylinder in Figure 3.3, which rolls without slipping on a dry horizontal plane propelled by the horizontally applied force $\mathbf{f}_a$.

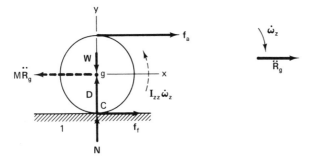

Figure 3.3   Rolling Cylinder

The moving xyz coordinate system is attached to the cylinder with its origin at g. The center of mass accelerates horizontally with $\ddot{\mathbf{R}}_g = \dot{\boldsymbol{\omega}}_z \times \mathbf{D}$. Positively sensed $\dot{\boldsymbol{\omega}}_z$ and $\ddot{\mathbf{R}}_g$ are shown to the right of the figure. The d'Alembert inertia force and couple are shown by dashed vectors acting on the free-body cylinder, contrary to the positive sense of the accelerations, as required by the d'Alembert formulation. The externally applied forces are also shown as acting on the cylinder just as they would be even if the cylinder did not accelerate. Friction force $\mathbf{f}_f$ is shown correctly since it opposes the cylinder's tendency to skid with a clockwise rotation.

Summing the forces that are shown acting on the free-body diagram yields the vector equation

$$\mathbf{N} - \mathbf{W} + \mathbf{f}_f + \mathbf{f}_a - M\ddot{\mathbf{R}}_g = 0$$

or, in component form,

$$N\mathbf{j} - W\mathbf{j} + f_f\mathbf{i} + f_a\mathbf{i} - M\ddot{R}_g\mathbf{i} = 0$$

which results in two scalar equations:

$$N - W = 0$$

$$f_f + f_a - M\ddot{R}_g = 0$$

For convenience, we will choose to take moments about the rolling contact point c because only the force $\mathbf{f}_a$ has a moment about c. Now summing moments and the couple $I_{zz}\dot{\boldsymbol{\omega}}_z$ about c,

$$2\mathbf{D} \times \mathbf{f}_a - \mathbf{D} \times M\ddot{\mathbf{R}}_g - I_{zz}\dot{\boldsymbol{\omega}}_z = \mathbf{0}$$

or, in component form, with $\ddot{\mathbf{R}}_g$ replaced by $\dot{\boldsymbol{\omega}}_z \times \mathbf{D}$,

$$2D\mathbf{j} \times f_a\mathbf{i} - D\mathbf{j} \times M(-\dot{\omega}_z\mathbf{k} \times D\mathbf{j}) + I_z\dot{\omega}_z\mathbf{k} = \mathbf{0}$$

or

$$-2Df_a\mathbf{k} + D^2M\dot{\omega}_z\mathbf{k} + I_z\dot{\omega}_z\mathbf{k} = \mathbf{0}$$

The resulting scalar equation is

$$-2Df_a + (D^2M + I_{zz})\dot{\omega}_z = 0$$

The coefficient $D^2M + I_{zz}$ is the moment of inertia the cylinder transferred from the center of mass (I_{zz}) to point c by D^2M. This is, of course, the "parallel-axis-theorem" transfer, which is usually covered in mechanics texts along with the study of moments and products of inertia. In this case, it is not necessary to introduce the parallel-axis theorem into the analysis. It appears naturally as the result of applying the d'Alembert formula.

The solution is now straightforward. Given f_a, the last equation can be solved for $\dot{\omega}_z$. Then with f_a and $\ddot{\mathbf{R}}_g$ calculated from $\dot{\boldsymbol{\omega}}_z \times \mathbf{D}$, the friction force f_f can be calculated. Of course, the problem is not "solved" until we verify that the Coulomb friction no-slip condition is satisfied, i.e.,

$$f_f \leq \mu_s N = \mu_s W$$

where μ_s is the static coefficient of friction. If this expression is satisfied, the solution is correct. If it is not, slip occurs and $\ddot{\mathbf{R}}_g \neq \dot{\boldsymbol{\omega}} \times \mathbf{D}$.

The solution with slip is also straightforward using the d'Alembert formulations. In that case,

$$f_f = \mu_d N = \mu_d W$$

where μ_d is the dynamic coefficient of friction, so that $M\ddot{\mathbf{R}}_g$ can be found immediately. The moment equation then becomes

$$2D\mathbf{j} \times f_a\mathbf{i} - D\mathbf{j} \times (f_a\mathbf{i} + f_f\mathbf{i}) + I_{zz}\dot{\omega}_z\mathbf{k} = \mathbf{0}$$

or

$$-Df_a + \mu_d WD + I_{zz}\dot{\omega}_z = 0$$

The analyst is always faced with the problem that there are two choices assignable to the direction of the friction. The correct direction is the one that opposes the tendency to slip, which is known when ω_z becomes known. With f_a and W known, two values of ω_z are found and the consistent result selected as the solution.

There are several important things to note about this solution. First, it was not necessary to anticipate the sense of either $\ddot{\mathbf{R}}_g$ or $\dot{\boldsymbol{\omega}}_z$, although that was easy to do in this relatively simple problem. In complicated problems, the best policy is to show both vectors in the positive sense of the coordinate system. Second, the d'Alembert inertia force and couple are always shown contrary to the positive sense of the linear and angular accelerations. Last, the sense and nature of the contact friction cannot always be anticipated, hence, special care must always be exercised to obtain a logically consistent solution.

3.5 CENTER OF PERCUSSION

The reason for introducing the rigid-body dynamic equations is because we intend to apply them to machine components. Before we do this, it would be advantageous to have them in their simplest form. To that end, we will now introduce the concept of the center of percussion.

Before proceeding with the development of this new concept, a few comments about the dynamic equations are in order. First, it should be noted that the linear momentum equation (Newton's second law) states that two vectors are equal, i.e., $\sum_{i=1}^{N} \mathbf{F}_i = M\ddot{\mathbf{R}}_g$. The vector $\ddot{\mathbf{R}}_g$ is obviously associated with the center of mass, but the second law does not require $\sum_{i=1}^{N} \mathbf{F}_i$ to pass through the center of mass. Figure 3.4 (which is not a free-body diagram) illustrates the general situation for a rigid body in which the required equality is satisfied. The summation of external forces $\sum_{i=1}^{N} \mathbf{F}_i$ can act anywhere on rigid body and the result will always be the same, i.e., $M\ddot{\mathbf{R}}_g$ will be parallel and equal in magnitude to $\sum_{i=1}^{N} \mathbf{F}_i$.

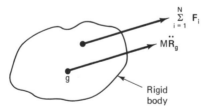

$$\sum_{i=1}^{N} \mathbf{F}_i$$

$$M\ddot{\mathbf{R}}_g$$

g

Rigid
body

Figure 3.4   Rigid Body Forces and
Inertia

Although it is not necessary to do so, as far as this equation is concerned, we usually think of $M\ddot{\mathbf{R}}_g$ as being applied at g because $\ddot{\mathbf{R}}_g$ is associated with that point. The reason $M\ddot{\mathbf{R}}_g$ must be shown passing through the center of mass of the rigid body on the free-body diagram is that the d'Alembert dynamic-moment equation envisions it at this location. The term $\mathbf{D} \times M\ddot{\mathbf{R}}_g$ places $M\ddot{\mathbf{R}}_g$ at the center of mass and computes the moment of this d'Alembert inertia force about point a.

With these ideas in mind, we are now going to ask, "what would happen if $M\ddot{\mathbf{R}}_g$ were attached to the free-body diagram at some location other than the center of mass?" As far as the second law is concerned (Newton's or d'Alembert's), nothing would happen. It might seem a little strange to show $M\ddot{\mathbf{R}}_g$ attached to a point that did not have the acceleration $\ddot{\mathbf{R}}_g$, but that would not change the way we would evaluate the summation of forces using the free-body diagram. However, the moment equation will not allow $M\ddot{\mathbf{R}}_g$ to be moved arbitrarily, since $\mathbf{D} \times M\ddot{\mathbf{R}}_g$ is a moment with a "bound" vector and not a couple with a "free" vector. The key word here is "arbitrarily" because $M\ddot{\mathbf{R}}_g$ could be moved parallel to itself if the change in its moment about a was compensated by the inclusion of a couple.

Figure 3.5 illustrates this process. Figure 3.5(a) shows the inertial force $M\ddot{\mathbf{R}}_g$ and the inertial couple $\dot{\mathbf{H}}_g$ as they would appear on a D'Alembert free-body diagram. In Figure 3.5(a), $M\ddot{\mathbf{R}}_g$ is shown acting through g, contrary to $\ddot{\mathbf{R}}_g$ and $\dot{\mathbf{H}}_g$, contrary to $\dot{\boldsymbol{\omega}}$. In Figure 3.5(b), a vector $M\ddot{\mathbf{R}}_g$ has been introduced at g and another vector $M\ddot{\mathbf{R}}_g$ attached at point cp to form a couple $\mathbf{e} \times M\ddot{\mathbf{R}}_g$. No net force has been added to the body by these vectors, but a clockwise couple has been introduced that was originally not there. Figures 3.5(a) and 3.5(b) are *not* equivalent because of the new couple. However, suppose that this new couple is equal to $\dot{\mathbf{H}}_g$. This could be arranged by properly selecting point cp. If that is done, then $\dot{\mathbf{H}}_g$ could be removed from the diagram, leaving $\mathbf{e} \times M\ddot{\mathbf{R}}_g$ to do its job.

Figure 3.5(c) shows a diagram equivalent to the one in Figure 3.5(a) when the couple $\mathbf{e} \times M\ddot{\mathbf{R}}_g = \dot{\mathbf{H}}_g$. The vector $\dot{\mathbf{H}}_g$ is missing from this diagram because $\mathbf{e} \times M\ddot{\mathbf{R}}_g$ takes its place. The two opposing forces acting at g can also be removed from this diagram. Figure 3.5(d) shows what remains when these forces are removed. It is equivalent to Figure 3.5(a), but simpler, since it contains only the d'Alembert force $M\ddot{\mathbf{R}}_g$ and not the d'Alembert couple $\dot{\mathbf{H}}_g$.

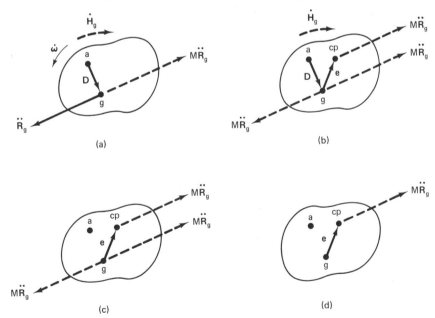

Figure 3.5   Center of Percussion

The vector **e** is a "shift vector" that locates cp relative to g. When $M\ddot{\mathbf{R}}_g$ and $\dot{\mathbf{H}}_g$ are known, it can be found from the equation

$$\mathbf{e} \times M\ddot{\mathbf{R}}_g = \dot{\mathbf{H}}_g \qquad (3.12)$$

Point cp is called the center of percussion for reasons that will become apparent in the next example. Let us return to the rolling cylinder example. Suppose that $\dot{\omega}_z$ is a known quantity in the clockwise direction. Then $\ddot{\mathbf{R}}_g$ is actually to the right in Figure 3.6 and can be calculated from $\dot{\boldsymbol{\omega}} \times \mathbf{D}$. The free-body inertial effects would then be as shown in Figure 3.6(a). The free-body diagram could be simplified to that shown in Figure 3.6(b) if the force $M\ddot{\mathbf{R}}_g$ were moved upward a distance $e = I_{zz}\dot{\omega}_z/M\ddot{R}_g$ so that its moment about g is exactly equal to the $I_{zz}\dot{\omega}_z$ it replaces. In the case of the rolling cylinder, where $\ddot{R}_g = \dot{\omega}_z D$ and $I_{zz} = MD^2/2$, the center of percussion is $e = D/2$, which is a fixed point in the cylinder independent of speed in this example. Unlike the center of mass, the location of the center of percussion is influenced by the motion of the body. For example, suppose the whole system (cylinder and ground) was accelerated to the right in Figure 3.3 with acceleration A_s without changing $\dot{\boldsymbol{\omega}}_z$. Then

$$e = \frac{I_{zz}\dot{\omega}_z}{M(\ddot{R}_g + A_s)} < \frac{D}{2}$$

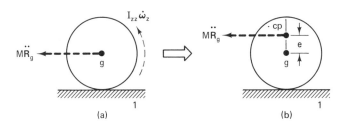

Figure 3.6   Rolling Cylinder Inertial Equivalency

A physical feel for the center of percussion can be obtained if we imagine the applied force to act at point cp, as shown in Figure 3.7. Taking moments about c shows that

$$(D + e)(f_a - M\ddot{R}_g) = 0$$

or

$$f_a - M\ddot{R}_g = 0$$

Application of the linear momentum equation yields

$$f_a + f_f - M\ddot{R}_g = 0$$

which means that $f_f = 0$. In this case, the cylinder would roll without the aid of friction, i.e., on a perfectly smooth surface. Most of us have experienced a similar phe-

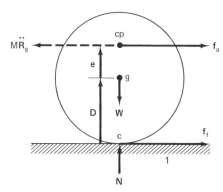

Figure 3.7   Rolling Cylinder Free Body

nomenon when we hit a ball with a bat or tennis racket in "just the right spot." We then have the sensation of achieving a long hit ball with very little effort.

3.6 RADIAL-AND-TRANSVERSE-FORCE-COMPONENT METHOD

From the preceding development, it is clear that the center of percussion is not generally a property of the solid, but depends on the instantaneous kinematics; in the planar case, it depends on the instantaneous ratio of kinematic quantities $\dot{\omega}_z$ and $\ddot{R}_g$. In the example cited, these quantities were proportional to each other, but that is not always the case. That restriction does not prove to be an important one if the dynamic forces and moments on a machine element are examined on an instant-by-instant basis. Considering our previous work with graphical kinematics, the center-of-percussion concept looks like it might be used to simplify graphical dynamic analysis. We shall see that this is true in the technique that will be described in the following paragraphs.

Figure 3.8 shows a planar link (3) separated from a system of linkages to form a free body. It will be assumed that the kinematic analysis of the linkage has already been done and that the accelerations of the link ends and its angular acceleration are already known. The first step in the graphical solution process requires finding the

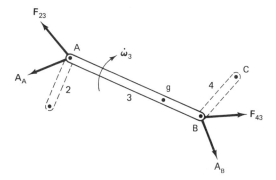

Figure 3.8   Planar Link

acceleration of the center of mass. It will be convenient to use $\mathbf{A}_g$ instead of $\ddot{\mathbf{R}}_g$ to denote the acceleration of the center of mass throughout the rest of this chapter. The acceleration $\mathbf{A}_g$ can be found graphically by drawing the link and $\mathbf{A}_A$ and $\mathbf{A}_B$ to scale, as shown in Figure 3.9.

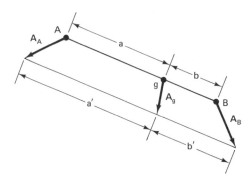

Figure 3.9   Graphical Solution for $\mathbf{A}_g$

Lengths a and b locating the center of mass are measured and scaled proportionately along the line joining the ends of vectors $\mathbf{A}_A$ and $\mathbf{A}_B$ according to

$$a'/a = b'/b$$

A vector drawn from g to the proportional point will be $\mathbf{A}_g$ to scale.

The proof of this "proportional" construction is obtained using the concepts developed in the previous chapter. The acceleration of g is found by adding radial $a\omega^2$ and transverse $a\dot{\omega}$ acceleration components to the acceleration of A_A, as shown in Figure 3.10. The length of the vector added to $\mathbf{A}_A$ is $a\sqrt{\dot{\omega}^2 + \omega^4}$.

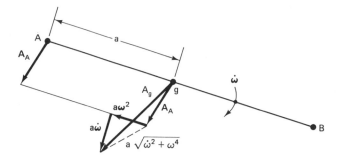

Figure 3.10   Acceleration $\mathbf{A}_g$

The acceleration of B can be found using the same method. The components added to $\mathbf{A}_A$ in this case are $(a + b)\omega^2$ and $(a + b)\dot{\omega}$, as shown in Figure 3.11. The length of the vector added to $\mathbf{A}_A$ in this case is $(a + b)\sqrt{\dot{\omega}^2 + \omega^4}$.

Figure 3.12 is extracted from Figure 3.11 to focus attention on two similar triangles, 1–2–3 and 1–4–5, which has triangle 1–2–3 superposed. Comparing these triangles shows that the terminus of $\mathbf{A}_g$ at point 3 divides line 1–5 joining the termini of $\mathbf{A}_A$ and $\mathbf{A}_B$ according to the proportion.

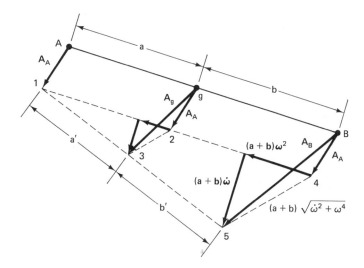

Figure 3.11   Acceleration $\mathbf{A}_B$

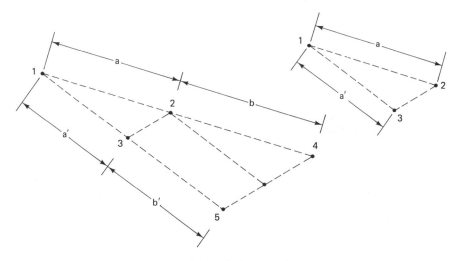

Figure 3.12   Similar Triangles

$$a'/a = b'/b = (a' + b')/(a + b)$$

The triangles are similar because the transverse and radial accelerations scale linearly with the distance from A. Obviously, the acceleration of any point on the line between A and B can be obtained using this proportional construction.

The acceleration of a point that is not on a line joining points A and B can be found using a similar procedure. Let g in Figure 3.13 be anywhere on a two-dimensional member. Assume that the accelerations of points A and B are known. To determine the acceleration of g, a construction line ab joining the ends of accel-

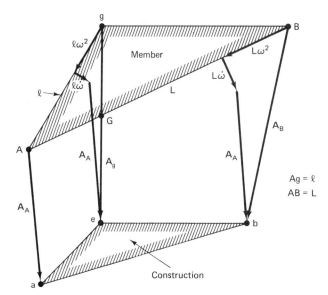

Figure 3.13   Acceleration of Point G

erations $\mathbf{A}_A$ and $\mathbf{A}_B$ is drawn as before. With ab as a base, triangle abe is constructed, which is similar to triangle ABg on the member, i.e.,

$$\frac{ae}{Ag} = \frac{ab}{AB} = \frac{eb}{gB}$$

The acceleration of g is the vector joining points g and e, as shown in Figure 3.13. Since the inertial force is transmissible, $\mathbf{A}_g$ can be attached to the member at G on line AB. The graphical analysis of the forces at A and B then proceeds as before. The proof of this construction given in the Appendix at the end of this chapter is based on similar triangles.

Next the d'Alembert inertia terms are shown acting on the link, as per Figure 3.14. They are shown opposite to $\dot{\boldsymbol{\omega}}_3$ and $\mathbf{A}_{g3}$, as required by the d'Alembert free-body diagram, and given names T_{o3} and F_{o3}.

The inertial couple is eliminated by calculating the ratio

$$e = \frac{I_{\omega 3}\dot{\omega}_3}{MA_{g3}} = \frac{T_o}{F_o} \tag{3.13}$$

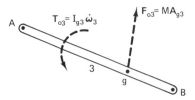

Figure 3.14   D'Alembert Inertia Terms

and shifting $\mathbf{MA}_{g3}$ parallel to itself in the direction that will cause a moment about g in the same direction as $I_{g3}\dot{\boldsymbol{\omega}}_3$. That construction is shown in Figure 3.15. (Be careful to measure e perpendicular to MA_{g3} not along member AB.)

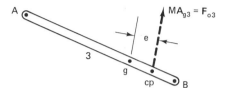

Figure 3.15   Transfer of $M\ddot{\mathbf{R}}_g$ to cp

This figure contains the entire d'Alembert inertial effect in the simplest way, i.e., in terms of a single force MA_{g3} passing through cp. We are now in a position to begin the analysis of the dynamic forces applied to the ends of links (3) by the adjacent links (2) and (4). Because all of the dynamic effects have been embodied in a single applied load (the inertia force $\mathbf{F}_{o3}$), the remaining forces required at A and B can be determined using the equilibrium equations of statics.

The "radial-and-transverse-force-component method" begins with the application of the dynamic-moment equation. The free-body diagram for the link is shown in Figure 3.16, which shows the inertia force and the force link (4) exerts on link (3) divided into radial and transverse components, $\mathbf{F}_{43}^{RA}$ and $\mathbf{F}_{43}^{TA}$.

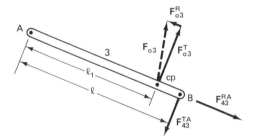

Figure 3.16   Radial and Transverse Force Components

Unfortunately, these components can only be roughly drawn since absolutely nothing is known about them at this point in the analysis. The subscripts on these components read "4 on 3" and the superscripts "radial about A" and "transverse about A." The inertia force has been abbreviated $\mathbf{F}_{03}$ and its components shown as $\mathbf{F}_{03}^{R}$, radial, and $\mathbf{F}_{03}^{T}$, transverse.

Remembering that the d'Alembert equations permit us to draw a free-body diagram and treat the problem as we would a static one, we can take moments about point A, stating

$$l_1 F_{03}^T = l F_{43}^{TA}$$

or

$$F_{43}^{TA} = \left(\frac{l_1}{l}\right) F_{03}^T$$

Since the right-hand side of this last equation contains all known quantities, the transverse component of the unknown force $\mathbf{F}_{43}$ could be calculated. Rather than do that, we will note that the ratio l_1/l simply scales $\mathbf{F}_{03}^T$ and then use a similar triangle construction to accomplish the same thing graphically. Figure 3.17 shows this construction. The component $\mathbf{F}_{03}^T$ is transferred to the B end of the link and a hypotenuse is drawn from its tip to A. The intersection (i) of the hypotenuse with the $\mathbf{F}_{03}^T$ emanating from cp is the magnitude of $\mathbf{F}_{34}^{TA}$. That magnitude is then transferred to the B end of link (3) and labeled $\mathbf{F}_{34}^{TA}$ because it is the transverse component that member (3) exerts on member (4), the opposite of what is shown in Figure 3.16.

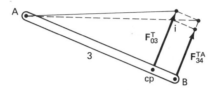

Figure 3.17 Similar Triangle Construction

This transverse component is all that can be found by applying the d'Alembert equations to member (3). However, if the same procedure were followed for member (4), a comparable component $\mathbf{F}_{34}^{TC}$ could also be found. Figure 3.18(a) shows the two components that can be found by taking moments about A and C. Let us return to the examination of member (3): both transverse components will be shown as they are applied to member (3) by (4), as shown in Figure 3.18(b).

The total force that (4) exerts on (3) should be the same whether its components come from an analysis of member (3) with A at one end or member (4) with C

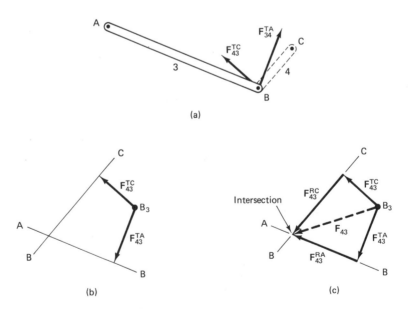

Figure 3.18 Force Analysis of Member (3)

at one end. Although the radial components $\mathbf{F}_{43}^{RA}$ and $\mathbf{F}_{43}^{RC}$ are of unknown magnitudes, their lines of action are known, i.e., parallel to their respective members. Figure 3.18(b) shows construction lines parallel to AB and BC drawn through the ends of the components $\mathbf{F}_{43}^{TA}$ and $\mathbf{F}_{43}^{TC}$. The missing radial components act along these lines. Figure 3.18(c) shows these radial components drawn along AB and BC, so that

$$\mathbf{F}_{43}^{TA} + \mathbf{F}_{43}^{RA} = \mathbf{F}_{43} = \mathbf{F}_{43}^{TC} + \mathbf{F}_{43}^{RC}$$

Closure is accomplished by the intersection of construction lines AB and BC. Superscripting with A and C helps label the radial components drawn along these lines.

Once the total force acting at one end of a member is found, the force at the other end can be found by envoking the yet unused d'Alembert force equation. Figure 3.19 shows the member with the now known forces $\mathbf{F}_{03}$ and $\mathbf{F}_{43}$ attached. These forces are shown in Figure 3.20 along with $\mathbf{F}_{23}$, which closes the triangle according to the d'Alembert force equation

$$\mathbf{F}_{43} + \mathbf{F}_{03} + \mathbf{F}_{23} = \mathbf{0}$$

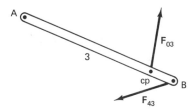

Figure 3.19 Force Analysis of Member (3)

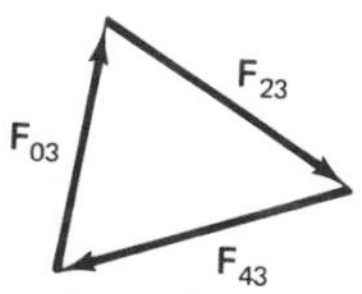

Figure 3.20
Force Analysis of
Member (3)

The analysis of all remaining members can now be performed in a like manner without further recourse to the radial-and-transverse-component method.

Very often one member of the linkage system will have an applied couple, known or unknown. When the magnitude and direction of that couple are known, it is best to start the analysis with that member, combining that couple with the inertia couple before computing e for that member. If the applied couple is unknown, it is best to leave that member to last, finding the unknown couple as the end result of the graphics. This procedure will be illustrated in examples to follow.

Scotch Yoke

Figure 3.21 shows a Scotch yoke similar to the one examined in the previous chapter. The results of a graphical kinematic analysis of the mechanism are shown in Figure 3.22. In the position shown, the instantaneous torque applied to the link (2) in the form of a couple is $\mathbf{T}_a$. The power associated with this torque is given by

$$\text{Power}_2 = \mathbf{T}_a \cdot \boldsymbol{\omega}_2$$

$$= -0.02(30) = -0.6 \text{ watt}$$

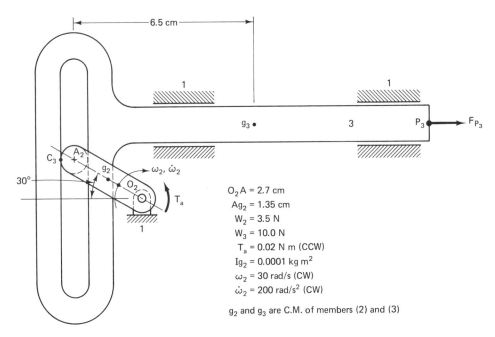

Figure 3.21   Scotch Yoke Mechanism

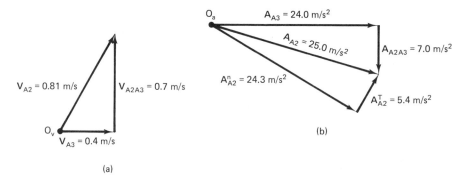

Figure 3.22   Velocity and Acceleration

The negative sign is the result of the opposite sense of the two vectors and indicates that power is being taken from the mechanism at point O_2.

For a conservative (frictionless) system without inertial effects, we would expect that power input at P_3, which is

$$\text{Power}_3 = \mathbf{F}_{P3} \cdot \mathbf{V}_{P3}$$

would be required to supply the 0.6 watt *plus* the rate at which work is done against gravity to increase the potential energy of the center of mass g_2 of link (2). Since real accelerating systems will have inertia effects, some of the input power at P_3 will also be used to change the rate at which kinetic energy is stored in crank (2) and yoke (3). Thus, even if the rate of change of the potential energy of link (2) were zero, as it would be if gravity were normal to the drawing, it would still not be correct to equate the input power to the output power.

In the next chapter, we will develop methods of writing "power equations" that include the effect of potential- and kinetic-energy rates of change. Unfortunately, these methods will tell us nothing about the forces and moments acting on individual members, which is our objective here. However, our observation about the power flow through the mechanism does give us an idea of what to expect and why.

Figure 3.23 shows link (2) with the accelerations of its ends drawn to scale and attached to each end ($A_{O2} = 0$). The dashed construction line joining O_2 and the tip of A_{A2} is bisected because g_2 is midway between A_2 and O_2. The acceleration A_{g2} is shown emanating from g_2, reaching to the point of bisection.

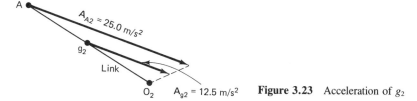

Figure 3.23   Acceleration of g_2

The d'Alembert inertial force is shown in Figure 3.24 as a dashed vector emanating from g_2, but contrary in sense to A_{g2}. The size of this vector is

$$F_{O2} = \frac{W_2}{g} A_{g2} = \left(\frac{3.5}{9.8}\right) 12.5 = 4.5 \text{ N}$$

The angular acceleration $\dot{\omega}_2$ shown in Figure 3.21 is clockwise so that the d'Alembert inertial couple is counterclockwise, which in this case is the same as T_a. The magnitude of this couple is

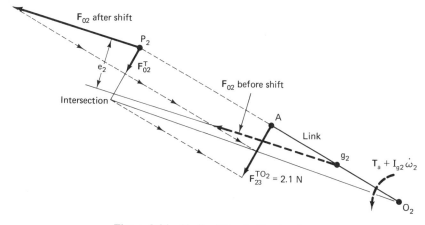

Figure 3.24   Similar Triangle Construction

$$T_{O2} = I_{g2}\dot{\omega}_2 = 0.0001(200) = 0.02 \text{ N-m}$$

A free-body diagram of this member would show these two couples acting together. To eliminate these couples from the free-body diagram we will shift the inertial d'Alembert force $\mathbf{F}_{O2}$ (shown dashed) parallel to itself so that its moment about g_2 is just equal to the combined effect of these two couples. The moment arm, or shift distance, is

$$e_2 = \frac{T_a + I_{g2}\dot{\omega}_2}{F_{O2}} = \frac{0.02 + 0.02}{4.5} = 0.009 \text{ m}$$

$$= 0.9 \text{ cm}$$

Figure 3.24 shows the d'Alembert force $\mathbf{F}_{O2}$ shifted through the distance e_2 to a new position on an extension of link (2) indicated by the dashed line. Notice that the moment of $\mathbf{F}_{O2}$ about g_2 is in the same sense as the couple $T_a + I_{g2}\dot{\omega}_2$, which is shown dashed. The point of intersection, P_2, is not the center of percussion. (The location of cp is determined by the inertial state of the member from $I_{g2}\dot{\omega}_2/F_{O2}$.)

The intersection P_2 of the shifted inertial force with the extension of link (2) does not mean that the similar-triangle technique cannot be used; it just requires a little more care in its use. Figure 3.24 shows this construction. The normal, or

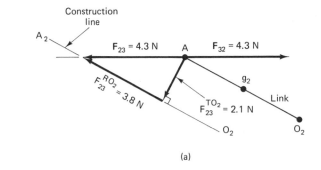

(a)

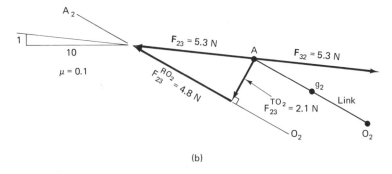

(b)

Figure 3.25   Radial and Transverse Components of $\mathbf{F}_{23}$

transverse, component of $\mathbf{F}_{O2}$ (called $(\mathbf{F}_{O2}^T)$ is erected at A and the usual hypotenuse drawn through its tip and O_2. In this case, however, the hypotenuse must be extended beyond the tip until it intersects a construction line that is normal to the extension of link (2) at P_2. A vector of that length is shown attached to point A at the end of the link. This is the transverse component of the force that link (2) exerts on link (3). The magnitude of this component is $F_{23}^{TO2} = 2.1$ N. This is as much as can be accomplished without considering the adjacent contacting member (3). At this point, we will neglect the effects of the dead-weight load of 3.5 N. Its effects will be introduced following the dynamic analysis.

If it is assumed that the contact between the crank and the yoke is frictionless, then the total force $\mathbf{F}_{23} = -\mathbf{F}_{32}$ at the contact and will be normal to the slot. Figure 3.25 shows the force components radial and transverse to the crank at the A end, which must sum to the force $\mathbf{F}_{23}$. Closure is obtained by intersecting the construction line $A_2 O_2$ and the horizontal line through A. Force $\mathbf{F}_{32}$ is also shown in Figure 3.25.

The force $\mathbf{F}_{23}$ that crank (2) exerts on yoke (3) at C_3 is shown in Figure 3.26 along with $\mathbf{F}_{O3}$. According to d'Alembert's version of Newton's second law, the summation of forces on the yoke is

$$\mathbf{F}_{23} + \mathbf{F}_{P3} + \mathbf{F}_{O3} = \mathbf{0}$$

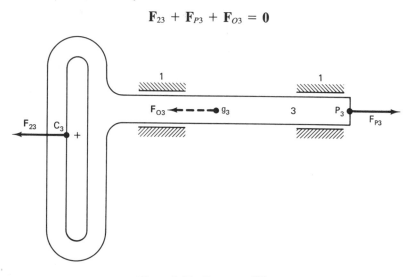

Figure 3.26   Forces on Yoke

The d'Alembert inertia force is

$$F_{O3} = 10(24)/9.8 = 24.5 \text{ N}$$

The force F_{P3} required to maintain the prescribed motion is

$$F_{P3} = F_{O3} + F_{23} = 24.5 + 4.3 = 28.8 \text{ N (to the right)}$$

The power input required at P_3 is

$$\text{Power}_3 = \mathbf{F}_{P3} \cdot \mathbf{V}_3 = (28.8)(0.4) = 11.5 \text{ watts}$$

The difference between the input 11.5 watts and the output 0.6 watts is due to the power required to cause a time rate of change of the kinetic energy of the crank and yoke. This power is not lost from the system. It is simply stored in the kinetic energy of the parts in a flywheellike effect. When the system comes to rest, all the energy stored kinematically will be restored to the surroundings. Meanwhile, these relatively massive parts and their accelerations are absorbing and returning power to the system's terminals.

This completes the analysis of the dynamic forces. The forces contributed by the dead weight can be obtained using the technique illustrated before. In this case, $\mathbf{F}_{O2}$ is the crank weight, 3.5 N, which acts vertically downward through the center of mass g_2, as shown in Figure 3.27(a). In this figure, $\mathbf{f}_{23}^{TO2}$ and $\mathbf{t}_s$ are the static force and torques, respectively, required to maintain the system at rest. These are in addition to the dynamic quantities $\mathbf{F}_{23}^{TO2}$ and $\mathbf{T}_s$ needed to attain the prescribed motion. The system can be maintained in equilibrium by applying torque t_s to the crank at O_2, or a horizontal force to yoke (3), or a combination of both. To illustrate the technique, it will be assumed that $\mathbf{t}_s$ is zero, so that the supporting force must be applied to tee-shaped member (3).

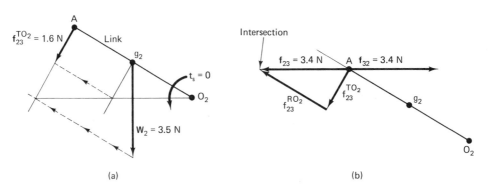

(a)   (b)

Figure 3.27   Static Analysis of Scotch Yoke

With $\mathbf{t}_s = 0$, the transverse force component $\mathbf{f}_{23}^{TO2}$ can be determined by the straightforward application of the similar-triangle technique, as illustrated in Figure 3.27(a). Closure for the frictionless case is indicated in Figure 3.27(b). To maintain equilibrium, a static force $f_{P3} = 3.4$ N is required at P_3. The total force at P_3 is $F_{P3} + f_{P3} = 32.2$ N and the total power is $(\mathbf{F}_{P3} + \mathbf{f}_{P3}) \cdot \mathbf{V}_3 = 12.9$ watts. The extra power $\mathbf{f}_{P3} \cdot \mathbf{V}_3$ produces the rate of change of the potential energy of the crank.

The presence of friction at c_3 alters the sense of the force $\mathbf{F}_{23}$ in Figure 3.25(a). The force that the crank exerts on the yoke now has a component normal to the yoke slot and an upward tangential friction component as well. With Coulomb or dry friction, the ratio of the friction force to the normal force equals the coefficient of friction. The coefficient of friction, therefore, establishes the direction of the total force relative to the surface, perpendicular if friction is zero and inclined when it is not.

Figure 3.25(b) shows the sense of $\mathbf{F}_{23}$ when the coefficient of friction is 0.1. The intersection of this construction line with the construction line for $\mathbf{F}_{23}^{RO2}$ yields a larger horizontal component of $\mathbf{F}_{23}$, i.e., 5.3 N. When the forces are summed on the yoke, friction increases F_{P3} to 29.8 N and the dynamic power to 11.9 watts.

Planetary-Gear System

There are occasions when the radial-and-transverse-component method cannot be implemented using the techniques just described. An obvious example is when two parallel links join together. In that case, it is impossible to intersect two radial components, as shown in Figure 3.18(c). A slight modification of the technique is required in these situations.

An example of a device requiring special care is shown in Figure 3.28. The acceleration diagram for points g_2 and g_3 is shown in Figure 3.29.

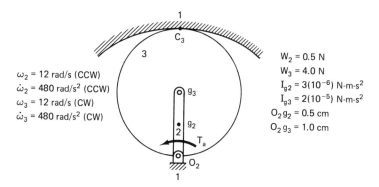

$\omega_2 = 12$ rad/s (CCW)
$\dot{\omega}_2 = 480$ rad/s^2 (CCW)
$\omega_3 = 12$ rad/s (CW)
$\dot{\omega}_3 = 480$ rad/s^2 (CW)

$W_2 = 0.5$ N
$W_3 = 4.0$ N
$I_{g2} = 3(10^{-6})$ N-m-s^2
$I_{g3} = 2(10^{-5})$ N-m-s^2
$O_2 g_2 = 0.5$ cm
$O_2 g_3 = 1.0$ cm

Figure 3.28   Planetary Gear System

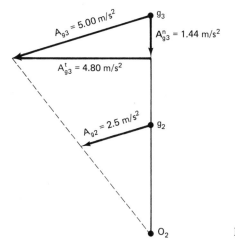

$A_{g3} = 5.00$ m/s^2

g_3

$A_{g3}^n = 1.44$ m/s^2

$A_{g3}^t = 4.80$ m/s^2

$A_{g2} = 2.5$ m/s^2 g_2

O_2

Figure 3.29   Accelerations of g_2 and g_3

We will begin the dynamic analysis with gear (3) since it is torque-free. Figure 3.30 shows the shifted inertia force $\mathbf{F}_{O3}$ and $\mathbf{F}_{32}^{TC3}$ obtained by triangulation. To make the system statically determinant, we will assume that no radial force is transmitted across the interface between C_3 and C_1. Of course, a transverse force will be required there to cause gear (3) to roll around the internal gear (1).

Figure 3.31 is a free-body diagram of member (3). The net moment of $\mathbf{F}_{23}^{TC3}$ $(= -\mathbf{F}_{32}^{TC3})$ and $\mathbf{F}_{O3}^{T}$ about C_3 is zero from the construction of Figure 3.30. In the absence of a radial force at C_3, $\mathbf{F}_{O3}^{R}$ must cancel $\mathbf{F}_{23}^{RC3}$, so that the sum of the radial-force components is zero. For the sum of the transverse-force components to vanish, $F_{13}^{T} = F_{23}^{TC3} - F_{O3}^{T} = 1.1$ N.

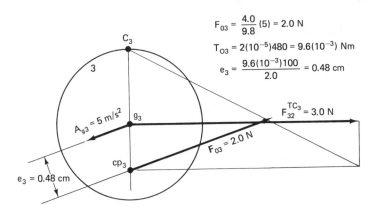

$$F_{03} = \frac{4.0}{9.8}(5) = 2.0 \text{ N}$$

$$T_{03} = 2(10^{-5})480 = 9.6(10^{-3}) \text{ Nm}$$

$$e_3 = \frac{9.6(10^{-3})100}{2.0} = 0.48 \text{ cm}$$

Figure 3.30 Transverse Force $F_{32}{}^{TC3}$

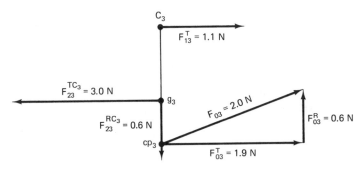

Figure 3.31 Freebody Diagram of Gear (3)

Figure 3.32 shows the location of F_{O2} relative to link (2). Since the applied torque $\mathbf{T}_{a2}$ is unknown, $\mathbf{F}_{32}^{TO2}$ cannot be established by triangulation on this diagram.

Figure 3.33 is a free body diagram of member (2). For the summation of radial- and transverse-force components to vanish

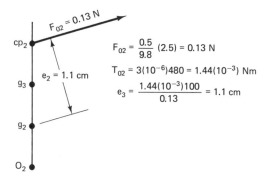

$$F_{02} = \frac{0.5}{9.8}(2.5) = 0.13 \text{ N}$$

$$T_{02} = 3(10^{-6})480 = 1.44(10^{-3}) \text{ Nm}$$

$$e_3 = \frac{1.44(10^{-3})100}{0.13} = 1.1 \text{ cm}$$

Figure 3.32 Location of CP_2

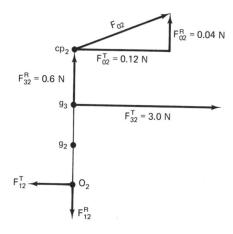

Figure 3.33 Freebody of Link (2)

$$F_{12}^R = F_{32}^R + F_{02}^R = 0.6 + 0.04 = 0.64 \text{ N}$$

$$F_{12}^T = F_{32}^T + F_{02}^T = 3.0 + 0.12 = 3.12 \text{ N}$$

However, the moment about O_2 is zero only if a couple or torque is applied to link (2).

The magnitude of this torque is

$$T_a = (O_2 g_3)F_{32}^T + (O_2 cp)F_{02}^T = 1(3.0) + 1.65(0.12) = 3.2 \text{ N-cm}$$

This is the torque required to accelerate the two components of the mechanism at the prescribed rate.

Notice that no triangulation was performed on $\mathbf{F}_{02}$ since $\mathbf{F}_{32}^T$ is already known from Newton's third law of action and reaction. The power input required to change the kinetic energy of the system is

$$\text{Power in} = T_a \omega_2$$

$$= 3.2(10^{-2})12$$

$$= 0.38 \text{ watts}$$

3.7 ELASTODYNAMIC ANALYSIS

The graphical methods described before yield the forces and couples applied to the connected members at the points where they are joined together. For simple beam-like members, the technique can also be used to determine the stresses and deflections within a member.

Figure 3.34 shows a link with known end forces $\mathbf{F}_{23}$ and $\mathbf{F}_{43}$ acting at A and B, respectively. We would like to determine the state of stress at a cross section of the link located at a. To do this, we will imagine that the lower portion of link (II) as a free body with a moment $\mathbf{M}_a$ and a force $\mathbf{F}_a$ applied to its cut-through cross section at a by the rest of member (I). To determine the inertial contributions, we must locate the center of mass of II, i.e., g_{II}, and find the acceleration of this point. This can be done using the graphical technique derived earlier. Then the inertial force $\mathbf{F}_{O\mathrm{II}} = m_{\mathrm{II}}\mathbf{A}_{g\mathrm{II}}$ can be determined and shown at g_{II} acting contrary to $\mathbf{A}_{g\mathrm{II}}$.

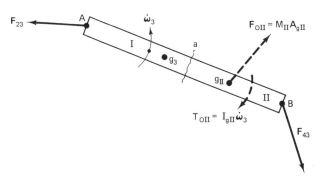

Figure 3.34 Forces on Link 3

To find the force $\mathbf{F}_a$ acting on the cross section at a, it is not necessary to shift $\mathbf{F}_a$ to the center of percussion of II since $\mathbf{F}_a$ is found by summing the forces according to the free-body diagram shown in Figure 3.35(a). The summation is shown in Figure 3.35(b).

The moment applied to II at a by I is obtained by summing moments about a, i.e.,

$$\mathbf{M}_a + \mathit{l}_{\mathrm{II}} \times \mathbf{F}_{43} + \mathbf{r} \times \mathbf{F}_{O\mathrm{II}} + \mathbf{T}_{O\mathrm{II}} = \mathbf{0}$$

The moments $\mathit{l}_{\mathrm{II}} \times \mathbf{F}_{43}$ and $\mathbf{r} \times \mathbf{F}_{O\mathrm{II}}$ can be found rather easily if the transverse components $\mathbf{F}_{43}^{TA}$ and $\mathbf{F}_{O\mathrm{II}}^{T}$ are first obtained graphically.

The transverse component of $\mathbf{F}_a$ determines the shear stress at a, and the radial component determines the tensile or compressive component of the axial stress. The bending component of the axial stress is found using M_a in the familiar formula $\sigma_m = M_a/Z$, where Z is the section modulus. The deflection analysis of the link due to distributed transverse and axial inertial forces and a distributed inertial couple can also be performed using the methods developed in strength-of-materials texts for beams and columns once the distribution of these loads is established. This will generally be a much more laborious task than the rather straightforward determination

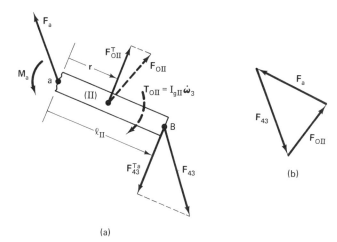

(a)

(b)

Figure 3.35 Free Body and Force Polygon

of the stresses. The important point to remember is that all the information required to perform these analyses can be obtained by applying the concepts developed earlier in this chapter.

To illustrate how the stresses at a cross section can be found, we will return to the Scotch-yoke mechanism of Figure 3.21 investigated earlier. Crank (2) will be assumed to be roughly a rectangular bar with a uniform cross section. To find the stresses on the cross section in the middle of the crank at g_2, we will need the acceleration of g_{II} midway between g_2 and O_2. Figure 3.36 shows the construction for A_{gII}. The moment of inertia of (II) can be scaled from I_{g2} by recalling that the moment of inertia for a bar is proportional to the mass of the bar and its length squared. Thus, $I_{gII} = \frac{1}{8}I_{g2} = 1.25(10^{-5})$ N-m and the d'Alembert couple is

$$T_{OII} = I_{gII}\dot{\omega}_2 = 0.0025 \text{ N-m (CCW)}$$

The d'Alembert force is

$$F_{OII} = \left(\frac{3.5}{2}\right)\left(\frac{6.25}{9.8}\right) = 1.1 \text{ N}$$

To find the force acting on the cross section at g_2, we must first find the force $\mathbf{F}_{12}$ that "ground" exerts on member (2). Force $\mathbf{F}_{12}$ is found from the force balance on the entire member (2), which is

$$\mathbf{F}_{32} + \mathbf{F}_{O2} + \mathbf{F}_{12} = 0$$

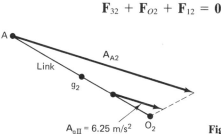

Figure 3.36 Acceleration $\mathbf{A}_g$

Vectors $\mathbf{F}_{O2}$ and $\mathbf{F}_{32}$ are obtained from Figures 3.24 and 3.25, respectively. The summation of forces that yield $\mathbf{F}_{12}$ is shown in Figure 3.37(a). The summation of forces acting on (II) is shown in Figure 3.37(b). A construction line parallel to O_2A_2, the center line of the crank, is also shown along with the radial and trans-components of $\mathbf{F}_{c2}$. The transverse component $\mathbf{F}_{g2}^T$ results in a shear stress. The radial component $\mathbf{F}_{g2}^R$ contributes a component to the axial stress.

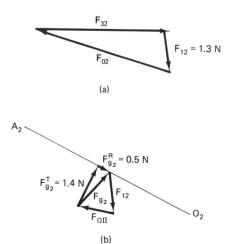

(a)

(b)   **Figure 3.37**   Force at Cross Section g_2

Bending of the crank also contributes to the axial stress. Figure 3.38 shows portion (II) of the crank (enlarged), with the applied forces and couples indicated. The transverse components of $\mathbf{F}_{OII}$ and $\mathbf{F}_{12}$ are also shown.

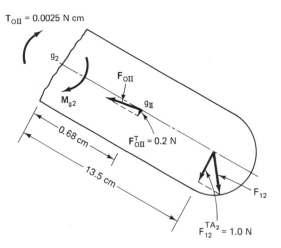

Figure 3.38   Moments about g_2

Summing the moments and couples yield

$$M_{g2} - 2.25 + (0.68)(0.2) + 1.35(1.0) = 0$$

$$M_{g2} = -0.76 \text{ N-cm}$$

The effect of $\mathbf{F}_{oII}$ is small so that $\mathbf{F}_{12}$ and $\mathbf{T}_a + \mathbf{T}_{oII}$ largely determine the bending stress at g_2. However, we should not overlook the fact that $\mathbf{F}_{12}$ is strongly influenced by the inertial force $\mathbf{F}_{o2}$, as indicated in Figure 3.37(a), so that dynamic forces do influence the stress levels.

PROBLEMS

It is intended that the following exercise problems be solved using the graphical methods developed in this chapter. To facilitate their solution it is suggested that the figures be photocopied and attached to the top of the solution page. All constructions can then be referenced to the geometry of the mechanism. All forces and torques are dynamic.

3.1. Determine the dynamic forces $\mathbf{F}_{14}$ and $\mathbf{F}_{12}$ acting on the mechanism shown in the diagram. Also find the shaft torque $\mathbf{T}_s$ applied to link (2) at O_2. Assume that link (2) and slider (3) are massless and I_{g4} is negligible. Use graphical methods.

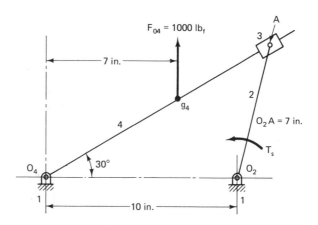

3.2. Block (4), which weighs 3 lb$_f$, is pulled by force $F_4 = 2$ lb$_f$ to the left, as shown in the diagram. Link (2) is massless, but link (3) is not. The acceleration diagram for the mechanism is also shown. Find the dynamic $\mathbf{F}_{34}$, neglecting friction between ground (1) and the block (4). Use graphical methods.

$W_3 = 4.0$ lb$_f$ $O_2A = 1.2$ in.
$AB = 2.8$ in. $g_3A = 1.4$ in.
$I_{g3} = 0.0006$ slug-ft^2 $W_4 = 3.0$ lb_f
$F_4 = 2.0$ lb$_f$

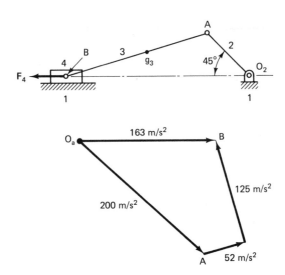

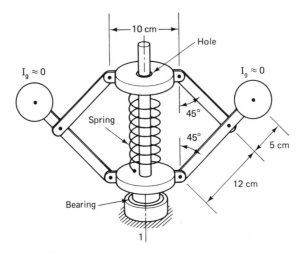

3.3 The flyball governor shown in the diagram rotates about the *Y–Y* axis at a constant angular velocity. The spring exerts a force of 100 lb$_f$ to balance the dynamic (inertial) effect on the balls. Each ball weighs 3.22 lb$_f$. Find the angular velocity of the governor. Solve by graphical methods.

3.4. A baseball player knows that if he hits a ball in just the right spot, the sensation of achieving a long hit ball with little effort occurs. Using the diagram, examine this problem and determine the location of the center of percussion.

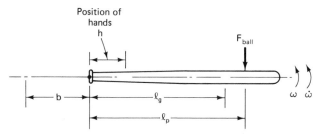

b = distance from center of rotation
ℓ_g = centroidal distance
ℓ_p = center of percussion
h = center of hands on bat

3.5. The diagram shows two links of a mechanism. The acceleration of the pin joining the members at A is also shown. The analysis of member (3) has been completed and the result, $\mathbf{F}_{32}^{TB} = 0.5$ lb$_f$, is shown. Using graphical methods, determine the vector $\mathbf{F}_{12}$ applied to member (2) by the ground (1) at O_2. Couple $\mathbf{T}_a(t)$ is applied to link (2) at O_2, as shown.

$O_2g_2 = 1.0$ in.	$O_2A = 1.5$ in.
$AB = 1.0$ in.	$A_A = 10.0$ in./s^2
$W_2 = 16.0$ lb$_f$	$I_{g2} = 1.3$ lb$_f$-in.-s^2
$T_{a(t)} = 0.14$ lb$_f$-in.	

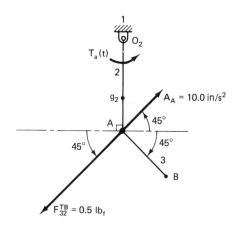

3.6. The diagram shows a 16-lb$_f$ cylinder rolling between two parallel accelerating surfaces. The coefficient of friction between the cylinder and these surfaces is $\mu = 0.2$. Assume that there is no slip at C_2 and C_4. Use graphical methods to determine the minimum normal clamping force N that must be applied to guarantee no slip. Neglect gravity effects. $W_3 = 16$ lb$_f$ and $I_{g3} = \frac{1}{4}$ lb$_f$-ft-s^2.

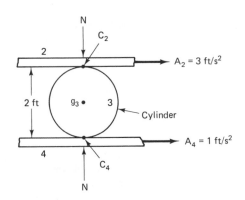

Plan view

3.7. Link (2) is released from rest ($\omega = 0$) when in the horizontal position shown in the diagram. Gravity acts downward to accelerate the link from rest. $I_{g2} = \frac{1}{12}mL^2$.

(a) Using graphical methods, determine $\dot{\omega}$ at time zero.

(b) Determine the reaction force at O_2 at time zero.

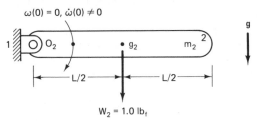

3.8. Links (2) and (4) are joined by a massless free-floating pin (3) that can slide in both slots, as shown in the diagram. Using graphical methods, determine the frictional forces that must be exerted on the slots to produce the motions indicated.

$O_2 P_3 = 4.2$ ft $\qquad$ $W_{pin} = 0$
$O_4 P_3 = 6.0$ ft $\qquad$ $T_{a2} = T_{a4} = 0$
$O_4 g_4 = 3.0$ ft $\qquad$ $I_{g2} = 2.6$ lb$_f$-ft-s^2
$O_2 g_2 = 3.0$ ft $\qquad$ $I_{g4} = 6.7$ lb$_f$-ft-s^2
$W_2 = W_4 = 10$ lb$_f$

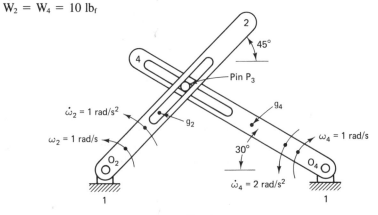

Plan view

3.9. Force *F* applied to block (2) causes it to slide in frictionless guideways with the velocity and acceleration shown in the diagram. Link (3) is driven by pin P_3 attached to the block, resulting in ω_3 and $\dot{\omega}_3$, as shown. The slot in member (3) is also frictionless. No external torque is applied to (3). Using graphical methods, find *F*.

$O_3g_3 = 2.5$ ft $O_3P_3 = 5.0$ ft
$m_2 = 5.0$ lbm $m_3 = 10.0$ lbm
$I_{g3} = 0.6$ lb$_f$-ft-s^2

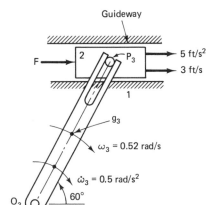

3.10. The slotted arm (2) shown in the diagram is used to drive the disk (3). Using graphical methods, determine the moment that the arm exerts on the disk through the frictionless slot and the pin on the rim of the disk.

$T_{a2} = 0.8\ (10^{-3})$ N-cm (CCW) $W_2 = 1.0$ N
$\omega_2 = 1.0$ rad/s (CCW) $I_{g2} = 1.0$ N $= 4.2\ (10^{-3})$ N-cm-s^2
$\dot{\omega}_2 = 0.5$ rad/s^2 (CCW)

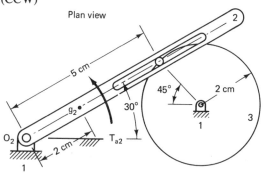

Diagram is to scale

3.11. A 10–lb$_f$ hollow cylinder (3) slides along rotating rod (2), as shown in the diagram. It touches the rod at *A* and *B*. Its acceleration at *A* in cylindrical coordinates is

$$\mathbf{A}_A = 3\mathbf{e}_r + 3\mathbf{e}_\theta\ \text{(ft/s}^2)$$

Its acceleration at *B* is

$$\mathbf{A}_B = 2\mathbf{e}_r + 4\mathbf{e}_\theta\ \text{(ft/s}^2)$$

Using graphical methods, find the normal forces on the rod at A and B.

$I_{g3} = 0.3$ lb$_f$-ft-s^2 $W_3 = 10.0$ lb$_f$.

$g_3B = 0.5$ ft

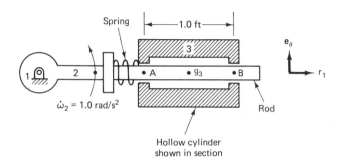

3.12. The diagram shows a slider–crank mechanism. The angular velocity and acceleration of (3) are shown. Using graphical methods, determine the dynamic torque T_s that must be applied to member (3) to obtain the motion indicated.

$O_3A = 4.0$ cm $O_3g_3 = 2.0$ cm

$W_3 = 0.5$ N $Ig_3 = 6.8 \, (10^{-6})$ N-m-s^2

$\omega_3 = 1.7$ rad/s (CCW) $\dot\omega_3 = 4.2$ rad/s^2 (CW)

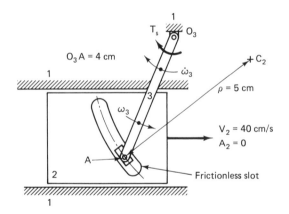

3.13. The diagram shows a slider–link mechanism to scale. Slider A has zero acceleration, but slider B accelerates at 1.6 ft/s^2. Force $F_B = 1.0$ lb$_f$ is applied to slider B opposite to its acceleration vector. The link (2) weighs 64.4 lb$_f$ and has $Ig_2 = 0.2$ lb$_f$-ft-s^2 about its center of mass, which is in the middle of the link. Find the normal force exerted by the slot on mass B for the frictionless case. $AB = 1.0$ ft. Acceleration A_B is not drawn to scale.

$W_2 = 64.4$ lb$_f$ $W_B = 32.2$ lb$_f$

$AB = 1.0$ ft $Ag_2 = 0.5$ ft

$I_{g2} = 0.2$ lb$_f$-ft-s^2

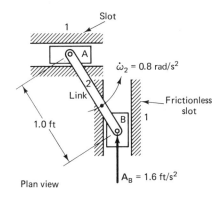

Plan view

3.14. For Problem 2.8 in Chapter 2, the data are as follows:

Disk:

$W_3 = 1000$ lb$_f$ $\dot{\omega}_3 = 0$

$I_{g3} = 30$ lb$_f$-ft-s^2 $T_a(t) = ?$

$\omega_3 = 2$ rad/s (CW)

Slider:

$W_s = 10$ lb$_f$ $I_{gs} = 0.002$ lb$_f$-ft-s^2

$O_3S = 1.0$ ft

Rod:

$W_2 = 2\,5$ lb$_f$ $\dot{\omega}_2 = 7.5$ rad/s^2 (CCW)

$I_{g2} = 0.3$ lb$_f$-ft-s^2 $O_2g_2 = \frac{1}{2}O_2S$

$\omega_2 = 2.1$ rad/s (CW)

Using graphical methods, determine the torque that much be applied to the shaft supporting the disk at O_3 or the torque applied by the shaft if that is the case assuming that the slot is frictionless.

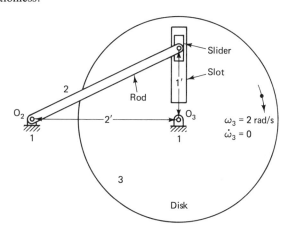

3.15. Link (3) of Problem 3.2 is a uniform beam element. Using the results of Problem 3.2, determine the force and moment acting on the cross section of this beam at g_3 using graphical methods.

3.16. Link (2) is caused to undergo motion so that the acceleration of end a is $A_a = 3$ m/s²
and of end b is 2 m/s². See the diagram.

   (a) Assuming that there is an external force at end a of 3.75 N directed along the slot
as shown and roller a is frictionless and weightless, what is the normal force N_a
acting on roller a? The slender rod is steel ($\rho_{steel} = 7,700$ kg/m³). The moment of
inertia of the square rod (each side is 1 cm) about its center of gravity is $\frac{1}{12}mL^2$.

   (b) Assuming that pin b is sliding with dry friction, determine the coefficient of friction μ and the direction of velocity b with respect to the ground.

$$ab = 9.7 \text{ m} \qquad A_a = 3 \text{ m/s}^2 \qquad L_2 = 10 \text{ m}$$
$$F_a = 3.75 \text{ N} \qquad A_b = 2 \text{ m/s}^2$$

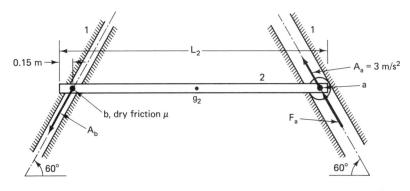

3.17. A portion of a mechanism is shown in the diagram; link (4) is shown cut through. The
guideway for the slider block (3) is frictionless. $W_3 = 0.06$ lb$_f$, $W_2 = 0.28$ lb$_f$, and
$I_{g2} = 1.0 \,(10^{-5})$ lb$_f$-ft-s².

   (a) Find $\dot{\omega}_2$.

   (b) Determine the total force of (2) on (3).

   (c) Determine the total force of (4) on (3).

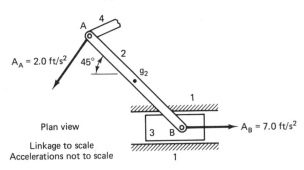

3.18. The diagram shows a cam-follower mechanism with an applied torque $\mathbf{T}_a$. Using graphical methods, determine the force exerted by (2) on (3) at c. Assume the contact is
frictionless. Neglect the effects of gravity.

$$O_2g_2 = 3.0 \text{ cm}$$
$$I_{g2} = 3.0 \,(10^{-6}) \text{ N-m-s}^2$$
$$W_2 = 1.0 \text{ N}$$

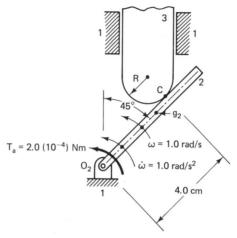

3.19. Planetary gear (3) is caused to rotate in contact with stationary gear (1) by applied torque $T_a = 0.02$ lb$_f$-in. applied to the massless carrier link (2) as shown in the diagram. Gear (4) is driven by gear (3). Using graphical methods determine the transverse force which gear (3) exerts on gear (4) at c.

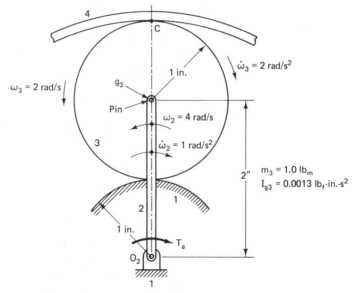

3.20. Slider block (3) is accelerated by force $\mathbf{F}_a$, as shown in the diagram. Rod (2), vertical at this instant, has the ω_2 and $\dot{\omega}_2$ shown at this instant. Using graphical methods, find the force F_a required to cause this motion. Assume frictionless motion. The upper end of (2) is free.

$W_2 = 32.2$ lb$_f$

$W_3 = 16.1$ lb$_f$

$I_{g2} = \frac{1}{12}mL^2$

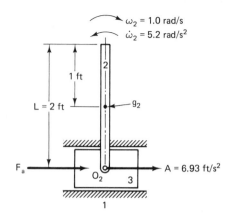

3.21. The diagram shows a link with its end accelerations attached. We would like to determine the inertially generated stress within the link at g, its center of gravity. To do this, we create a free body by cutting through the link at g and isolate the left portion as a free body.

$\mathbf{F}_a = 8\,(10^{-4})$ N

$\mathbf{m_I} = 3.63\,(10^{-3})$ kg

$\mathbf{L_I} = L_{II} = 2.4$ cm

$\mathbf{ag_I} = bg_{II} = 1.2$ cm

Using graphical methods, determine the following:

(a) Find $\mathbf{F}_o$ for this free body.

(b) Find $\mathbf{I}_g$ for this free body.

(c) Shift $\mathbf{F}_o$ to eliminate $I_g\dot{\omega}$.

(d) Find the radial and transverse components of the force at g.

(e) Find the bending moment at g.

Is the member in tension or compression at g due to inertial effects?

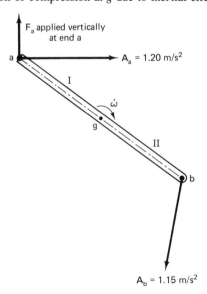

REFERENCES

1. Shames, I. H. *Engineering Mechanics—Statics and Dynamics,* 3rd Ed. Englewood Cliffs, NJ: Prentice Hall, 1980.
2. Mabie, H. H., and Ocvirk, F. W. *Mechanisms and Dynamics of Machinery,* 3rd Ed. New York: John Wiley, 1975.

APPENDIX: Acceleration of a Point on a Two-Dimensional Member

The known accelerations at points A and B on the two-dimensional ABg member are shown in the diagram. The acceleration of g obtained by vector addition is shown attached to g. Triangles DgE and dBf are similar. As a consequence, triangles AEg and AfB are also similar. Triangles abe and AfE are identical since AfE is abe shifted by the length of $\mathbf{A}_A$.

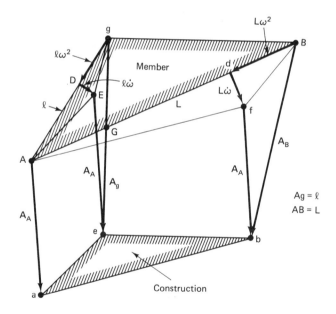

The proof is obtained from the similar triangles AEg and AfB where

$$\frac{Ag}{AB} = \frac{AE}{Af}$$

Since the angles gAB and EAf between these sides are equal

$$\frac{gB}{AB} = \frac{Ef}{Af}$$

which demonstrates that triangles ABg and abe are similar.

4

LAGRANGE
EQUATIONS

4.1 INTRODUCTION

In the previous chapter, the forces and couples acting on machine elements were found by following a three-step serial procedure. First, the accelerations of the elements were established starting with given kinematic inputs. Then the inertial effects were determined and applied to free-body diagrams using d'Alembert's technique. Element interconnecting loads were found last from the equations of equilibrium for each free body.

Unfortunately, not all dynamic problems can be solved following this procedure. As an example, consider the motion of a cam-follower system that has separated from the cam surface due to inertial effects. The vertical displacement of the follower, shown in Figure 4.1, is no longer related to the motion of the cam during separation. The time it takes for the follower to regain contact with the cam depends on the spring stiffness, the follower mass, and the kinematics of the follower when it left the cam. In this case, the dynamics control the kinematics, not the other way around, as in the previous chapter.

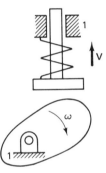

Figure 4.1 Cam-Follower System

The techniques of the previous chapter are not suitable for situations where the formulation and solution of the dynamics problem yield the kinematics of the mechanism. In this chapter, we will develop a method that permits the formulation of the kinematic equations for complex interconnected systems with a minimum amount of effort.

4.2 LAGRANGE EQUATIONS FOR A PARTICLE

Newton's second law for a particle used in the previous chapter is

$$\sum_{j=1}^{N} \mathbf{f}_{ij} + \mathbf{F}_i = \frac{d}{dt}(m_i \dot{\mathbf{r}}_{oi}) \tag{3.1}$$

To obtain the scalar equations of motion, let the vectors in this equation be written in component form, i.e.,

$$\sum_{j=1}^{N} \mathbf{f}_{ij} = \sum_{j=1}^{N} (f_{ij})_x \mathbf{I} + \sum_{j=1}^{N} (f_{ij})_y \mathbf{J} + \sum_{j=1}^{N} (f_{ij})_z \mathbf{K}$$

$$\mathbf{F}_i = (F_i)_x \mathbf{I} + (F_i)_y \mathbf{J} + (F_i)_z \mathbf{K}$$

$$\dot{\mathbf{r}}_{oi} = \dot{X}_i \mathbf{I} + \dot{Y}_i \mathbf{J} + \dot{Z}_i \mathbf{K}$$

The three equivalent scalar equations are then

$$\sum_{j=1}^{N} (f_{ij})_x + (F_i)_x = \frac{d}{dt}(m_i \dot{X}_i)$$

$$\sum_{j=1}^{N} (f_{ij})_y + (F_i)_y = \frac{d}{dt}(m_i \dot{Y}_i) \qquad (4.1)$$

$$\sum_{j=1}^{N} (f_{ij})_z + (F_i)_z = \frac{d}{dt}(m_i \dot{Z}_i)$$

The kinetic energy of a particle can be expressed in the form[†]

$$T_i = m_i/2(\dot{X}_i^2 + \dot{Y}_i^2 + \dot{Z}_i^2) \qquad (4.2)$$

Note that the components of the particle's momentum can be retrieved from the total kinetic energy by partial differentiating as follows:

$$\frac{\partial T_i}{\partial \dot{X}_i} = m_i \dot{X}_i \qquad \frac{\partial T_i}{\partial \dot{Y}_i} = m_i \dot{Y}_i \qquad \frac{\partial T_i}{\partial \dot{Z}_i} = m_i \dot{Z}_i$$

These expressions allow the components of the inertia on the right-hand side of Newton's equations to be replaced by

$$\frac{d}{dt}\left(\frac{\partial T_i}{\partial \dot{X}_i}\right) \qquad \frac{d}{dt}\left(\frac{\partial T_i}{\partial \dot{Y}_i}\right) \qquad \frac{d}{dt}\left(\frac{\partial T_i}{\partial \dot{Z}_i}\right)$$

The forces $\sum_{j=1}^{N} \mathbf{f}_{ij}$ on the left-hand side of Newton's equations are internal to a system of particles and will eventually be assumed to be self-cancelling for the collection of particles as a whole. The externally applied forces $\mathbf{F}_i$ can be divided into two general categories: those that can be obtained by taking the gradient of a potential and those that cannot. In the first category, we will be concerned with the earth's gravitational force and elastic forces. The second category consists of the "surface forces" due to contact with other bodies. Applied normal and tangential (friction) forces fall into this group.

Figure 4.2 shows a particle at an elevation Z above the surface of the earth. The potential energy of the particle relative to the earth's surface (zero potential) is

$$V_{gi} = m_i g Z$$

[†] In cylindrical coordinates, $T_i = m_i/2[\dot{r}^2 + (r\dot{\theta})^2 + \dot{z}^2]$

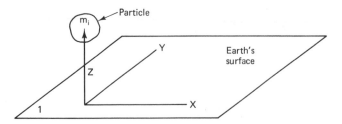

Figure 4.2 Potential of a Particle

The force that the earth exerts on this particle (for particles near the earth's surface) can be written as the negative gradient of this potential. Recall that the gradient operation is defined by

$$\nabla(\ ) = \mathbf{I}\frac{\partial(\ )}{\partial X} + \mathbf{J}\frac{\partial(\ )}{\partial Y} + \mathbf{K}\frac{\partial(\ )}{\partial Z}$$

so that

$$\mathbf{F}_{gi} = -\nabla V_{gi} = -m_i g\left(\mathbf{I}\frac{\partial Z}{\partial X} + \mathbf{J}\frac{\partial Z}{\partial Y} + \mathbf{K}\frac{\partial Z}{\partial Z}\right) = -m_i g\,\mathbf{K} \qquad (4.3)$$

which is the force of gravity.

Figure 4.3(a) shows a spring in its unloaded state with no energy stored elastically. Figure 4.3(b) shows the same spring stretched a distance X_i with an internal energy storage of

$$V_{si} = \tfrac{1}{2}kX_i^2$$

where k is the spring constant. The force applied to the free-body mass m_i by the spring is given by the negative gradient of the potential energy according to

$$\mathbf{F}_{si} = -\nabla V_{si} = -kX_i\mathbf{I}$$

In general, conservative forces can be expressed as the gradient of an appropriate potential so that

$$(F_i)_{\text{conservative}} = -\nabla V_i = -\mathbf{I}\frac{\partial V_i}{\partial X} - \mathbf{J}\frac{\partial V_i}{\partial X} - \mathbf{K}\frac{\partial V_i}{\partial Z}$$

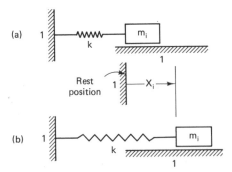

Figure 4.3 Energy Storage in a Spring

Unfortunately, nonconservative forces, such as friction, are usually not conveniently expressed as the gradient of a potential. For that reason, we will continue to treat them as external forces applied to the system. External forces are usually given the symbol Q, so that

$$\mathbf{Q}_i = (Q_i)_X \mathbf{I} + (Q_i)_Y \mathbf{J} + (Q_i)_Z \mathbf{K}$$

If the external forces, conservative and nonconservative, are written in the form

$$\mathbf{F}_i = -\nabla V_i + \mathbf{Q}_i \tag{4.4}$$

then the Lagrange equations of motion become

$$\sum_{j=1}^{N} (f_{ij})_X + (Q_i)_X = \frac{\partial V_i}{\partial X_i} + \frac{d}{dt}\left(\frac{\partial T_i}{\partial \dot{X}_i}\right)$$

$$\sum_{j=1}^{N} (f_{ij})_Y + (Q_i)_Y = \frac{\partial V_i}{\partial Y_i} + \frac{d}{dt}\left(\frac{\partial T_i}{\partial \dot{Y}_i}\right)$$

$$\sum_{j=1}^{N} (f_{ij})_Z + (Q_i)_Z = \frac{\partial V_i}{\partial Z_i} + \frac{d}{dt}\left(\frac{\partial T_i}{\partial \dot{Z}_i}\right)$$

4.3 LAGRANGE EQUATIONS FOR A RIGID BODY

As usual, we are not so much interested in the motion of the individual particles as their collective rigid-body motion. When the summation over all particles is performed, if $\mathbf{f}_{ij} = -\mathbf{f}_{ji}$, then the mutually attractive internal forces cancel. The sum of the external forces, Q_i, replaces the individual forces, and the individual potential energies become the potential energy of the body's center of mass. Summing the kinetic energies yields the total kinetic energy of the rigid body, T. For a system of particles comprising a single rigid body, the Lagrange equations of motion now become

$$Q_X = \frac{\partial V}{\partial X} + \frac{d}{dt}\left(\frac{\partial T}{\partial \dot{X}}\right)$$

$$Q_Y = \frac{\partial V}{\partial Y} + \frac{d}{dt}\left(\frac{\partial T}{\partial \dot{Y}}\right) \tag{4.5}$$

$$Q_Z = \frac{\partial V}{\partial Z} + \frac{d}{dt}\left(\frac{\partial T}{\partial \dot{Z}}\right)$$

Reviewing the formulations for V and T shows that V is a function of X, Y, and Z only, and T is a function only of the velocity components $\dot{X}$, $\dot{Y}$, and $\dot{Z}$. Mathematically, these dependencies are expressed by

$$V = V(X, Y, Z)$$

$$T = T(\dot{X}, \dot{Y}, \dot{Z})$$

For this reason, it is customary to combine T and V into a quantity called the Lagrangian, defined by

$$L = T - V \qquad (4.6)$$

Because of the dependencies noted before,

$$\frac{\partial L}{\partial \dot{X}} = \frac{\partial T}{\partial \dot{X}}, \text{ etc.}$$

and

$$\frac{\partial L}{\partial X} = -\frac{\partial V}{\partial X}, \text{ etc.}$$

The *rigid-body* Lagrange equations can be written in terms of the generalized coordinates X, Y, and Z, and their time derivatives as

$$\frac{d}{dt}\left(\frac{\partial L}{\partial \dot{X}}\right) - \frac{\partial L}{\partial X} = Q_X$$

$$\frac{d}{dt}\left(\frac{\partial L}{\partial \dot{Y}}\right) - \frac{\partial L}{\partial Y} = Q_Y \qquad (4.7)$$

$$\frac{d}{dt}\left(\frac{\partial L}{\partial \dot{Z}}\right) - \frac{\partial L}{\partial Z} = Q_Z$$

4.4 PHYSICAL INTERPRETATION OF THE LAGRANGE EQUATIONS

Before proceeding to generalize the Lagrange equations for a collection of interconnected rigid bodies such as occur in mechanisms, we will first pause to examine their physical interpretation. The dependent variable on the left-hand side of the equations is the total energy stored in the system. It may seem a little strange that V is subtracted from T rather than added, but that is simply a result of having expressed conservative forces as the negative gradient of a potential.

The potential and kinetic energies, and hence L, are expressed in terms of the state variables X, Y, Z, and $\dot{X}$, $\dot{Y}$, $\dot{Z}$, which locate and give the momentum of the body. The state variables are functions of time. Usually, only their initial values are known. The operations performed on L represent changes in the kinetic and potential energies, which, taken together are consistent with Newton's second law and the assumption that the internal forces are self-canceling.

The forces Q_X, Q_Y, and Q_Z are the external forces that cause the kinetic and potential energies to change. Forces that do not add or remove energy from the body are therefore excluded from the Q forces. These excludable forces are found among those that constrain the body's motion. Take, for example, the cylinder that rolls without slip, which was examined in the last chapter. The normal force N and the friction force f_f constrain the cylinder to roll on a flat surface. No energy is added to or removed from the cylinder by these constraining forces. If that were not true, we could remove the applied force f_a and allow f_f and N to propel the cylinder. Since spontaneous motion of this kind is not observed, we must conclude that these are "workless" constraining forces, which should not be included among the Q forces.

   If slip does occur, then energy is dissipated, i.e., removed from the system. This constraint is no longer workless, but is instead characterized by a Q force that opposes the motion and is therefore negative.

4.5 KINETIC ENERGY FOR A RIGID BODY

In Section 4.3, we said that a rigid body's kinetic energy is the sum of its particles' kinetic energies. We called this sum T without any indication of how this summing process might be performed. In this section, we will develop a formula for this sum that uses many of the same quantities that appear in the linear and angular momentum equations of the previous chapter.

   The total kinetic energy of a rigid body treated as a continuum is given by

$$T = \iiint (\dot{X}^2 + \dot{Y}^2 + \dot{Z}^2)\, dm$$

where the integration process replaces the summation over discrete particles.

   The integrand of this equation is just the dot product of the differential mass dm's velocity with itself. To verify this, let the velocity of dm be

$$\dot{\mathbf{r}}_o = \dot{X}\mathbf{I} + \dot{Y}\mathbf{J} + \dot{Z}\mathbf{K}$$

then

$$\dot{\mathbf{r}}_o \cdot \dot{\mathbf{r}}_o = \dot{X}^2 + \dot{Y}^2 + \dot{Z}^2$$

   In the past, it has always been more convenient to characterize the motion of a rigid body in terms of the motion of its center of mass and its angular motion rather than in terms of the motion of its individual particles.

   Figure 4.4 shows a moving coordinate system with its origin fixed at the center

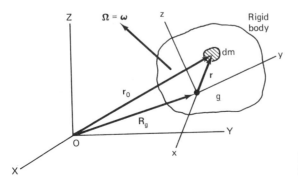

Figure 4.4 Moving Coordinates Fixed in a Rigid Body

of mass of the rigid body. This coordinate system also rotates with the body so that $\boldsymbol{\Omega} = \boldsymbol{\omega}$. In this system, the particles of a rigid body have no relative velocity, so that the velocity of dm is

$$\dot{\mathbf{r}}_o = \dot{\mathbf{R}}_g + \boldsymbol{\omega} \times \mathbf{r}$$

Forming the dot product yields

$$\dot{\mathbf{r}}_o \cdot \dot{\mathbf{r}}_o = \dot{\mathbf{R}}_g \cdot \dot{\mathbf{R}}_g + 2\dot{\mathbf{R}}_g \cdot (\boldsymbol{\omega} \times \mathbf{r}) + (\boldsymbol{\omega} \times \mathbf{r}) \cdot (\boldsymbol{\omega} \times \mathbf{r})$$

The total kinetic energy can now be expressed as

$$T = \tfrac{1}{2} \iiint \dot{\mathbf{R}}_g \cdot \dot{\mathbf{R}}_g \, dm + \tfrac{1}{2} \iiint 2\dot{\mathbf{R}}_g \cdot (\boldsymbol{\omega} \times \mathbf{r}) \, dm$$

$$+ \tfrac{1}{2} \iiint (\boldsymbol{\omega} \times \mathbf{r}) \cdot (\boldsymbol{\omega} \times \mathbf{r}) \, dm$$

Since the center of mass g is a fixed point in the body, the first integral becomes

$$\tfrac{1}{2} \iiint \dot{\mathbf{R}}_g \cdot \dot{\mathbf{R}}_g \, dm = \tfrac{1}{2} m (\dot{X}_g^2 + \dot{Y}_g^2 + \dot{Z}_g^2) = \tfrac{1}{2} m \dot{R}_g^2$$

which is the kinetic energy associated with the motion of the center of mass.

The integrand of the second integral varies only because r depends on the location of each dm. Vectors $\dot{\mathbf{R}}_g$ and $\boldsymbol{\omega}$ do not depend on the location of dm and, therefore, do not participate in the integration process. Consequently, this term can be written

$$\tfrac{1}{2} \iiint 2\dot{\mathbf{R}}_g \cdot (\boldsymbol{\omega} \times \mathbf{r}) \, dm = \dot{\mathbf{R}}_g \cdot \left(\boldsymbol{\omega} \times \iiint \mathbf{r} \, dm \right)$$

Recall that the definition of the center of mass is

$$\mathbf{r}_g = \frac{1}{m} \iiint \mathbf{r} \, dm$$

The quantity $\mathbf{r}_g = \mathbf{0}$ since the origin of xyz is at the center of mass. Locating xyz at g has proven to be a very convenient choice.

When the indicated operations of the last integral are performed, the result is

$$\iiint (\boldsymbol{\omega} \times \mathbf{r}) \cdot (\boldsymbol{\omega} \times \mathbf{r}) \, dm = I_{xx}\omega_x^2 + I_{yy}\omega_y^2 + I_{zz}\omega_z^2$$

$$- 2(\omega_x \omega_y I_{xy} + \omega_y \omega_z I_{yz} + \omega_z \omega_x I_{zx})$$

This group of terms represents the rotational kinetic-energy contribution of the body to the total kinetic energy.

The total kinetic energy of the body can now be expressed as the sum of translational and rotational kinetic energies, i.e.,

$$T = \tfrac{1}{2}m\dot{R}_g^2 + \tfrac{1}{2}[I_{xx}\dot{\theta}_x^2 + I_{yy}\dot{\theta}_y^2 + I_{zz}\dot{\theta}_z^2 - 2(\dot{\theta}_x\dot{\theta}_y I_{xy} + \dot{\theta}_y\dot{\theta}_z I_{yz} + \dot{\theta}_x\dot{\theta}_z I_{xz})]^{\dagger} \qquad (4.8)$$

In this formulation, the lowercase subscripts indicate that the moments and products of inertia are taken relative to the xyz coordinates whose origin is at the center of mass.

If X_g, Y_g, Z_g, and θ_x, θ_y, θ_z are independent, there will be six Lagrange equations, one for each of the six "degrees of freedom." They take the form

$$\frac{d}{dt}\left(\frac{dL}{d\dot{X}_g}\right) - \frac{dL}{dX_g} = Q_{xg}$$

$$\vdots$$

$$\frac{d}{dt}\left(\frac{\partial L}{\partial \dot{\theta}_x}\right) - \frac{\partial L}{\partial \theta_x} = Q_{\theta x} \qquad (4.9)$$

$$\vdots$$

In these equations Q_{xg}, Q_{yg}, and Q_{zg} are actually forces, but $Q_{\theta x}$, $Q_{\theta y}$, and $Q_{\theta z}$ are couples. Special care must be exercised in the assignment of the Q forces and couples. The product of each of these with an increment of its associated generalized coordinate must physically represent an increment of work done by that Q upon the system. In other words, the incremental work done by Q_{xg} must be $Q_{xg}\,\delta x_g$ and the incremental work done by Q_θ must be $Q_\theta\,d\theta$.

Coordinates and their derivatives are independent if they cannot be directly related to each other by kinematic expressions of the type used in Chapter 2. There is a Lagrange equation corresponding to each independent (generalized) coordinate in the Lagrangian. If the coordinates selected *are* dependent, the expression relating them can be substituted into the Lagrangian so it contains only independent (generalized) coordinates.

A single force such as shown in Figure 4.5(a) does work in a small displacement of

$$dW = \mathbf{Q}\cdot d\mathbf{s} = Q_x\,dx + Q_y\,dy + Q_z\,dz$$

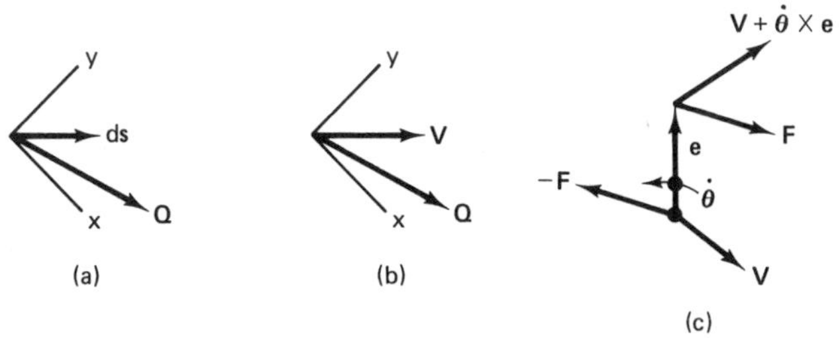

Figure 4.5 Generalized "Forces"

$^{\dagger}\omega_x = \dot{\theta}_x$, $\omega_y = \dot{\theta}_y$, and $\omega_z = \dot{\theta}_z$.

or delivers energy at a rate

$$dw/dt = \mathbf{Q} \cdot \mathbf{V} = Q_x \dot{x} + Q_y \dot{y} + Q_z \dot{z}$$

as indicated in Figure 4.5(b). The components of Q yield the Q_x, Q_y, and Q_z of the first three Lagrange equations if x, y, and z are the generalized coordinates.

On the other hand, the couple shown in Figure 4.5(c) delivers energy at a rate

$$dw/dt = -\mathbf{F} \cdot \mathbf{V} + \mathbf{F} \cdot (\mathbf{V} + \dot{\boldsymbol{\theta}} \times \mathbf{e}) = \mathbf{F} \cdot (\dot{\boldsymbol{\theta}} \times \mathbf{e}) = \dot{\boldsymbol{\theta}} \cdot (\mathbf{e} \times \mathbf{F})$$

where $\mathbf{e} \times \mathbf{F}$ is the couple vector. The components of $\mathbf{e} \times \mathbf{F}$ yield $Q_{\theta x}$, $Q_{\theta y}$, and $Q_{\theta z}$ of the second three Lagrange equations. To correctly identify these couples when they are not indicated as inputs, it will be necessary to reduce the force systems to "equivalent systems" of couples and "workless" constraint forces.

The Lagrange equations can be derived in a somewhat more elegant manner based on energy concepts. The connection with Newton's laws is not quite so evident when this approach is used. This alternative derivation is given in the appendix to this chapter for completeness and because it demonstrates that the formulation has its roots in energy concepts.

4.6 EXAMPLES OF THE APPLICATION OF LAGRANGE EQUATIONS TO SINGLE ELEMENT AND MULTIELEMENT SYSTEMS

The examples in this section have been chosen to illustrate certain features of the Lagrange-equation formulation that were discussed in the preceding section. Some new and interesting features will also be discovered.

Example 1: Vibrating Mass without Damping

Figure 4.6(a) shows a linear spring k whose original length is l_o.

When mass m is added, the spring stretches a length l to support its dead weight. When excitation $f_a(t)$ is applied to mass m, it displaces a distance X measured relative to the unstretched length of the spring. Since no energy is stored in the spring when X equals zero, we can write the potential energy of the system as

$$V = \tfrac{1}{2} kX^2 - mgX$$

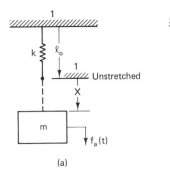

(a)

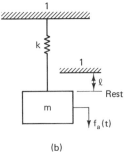

(b)

Figure 4.6 Vibrating Mass—No Damping

The Lagrangian for the mass–spring system is, therefore,

$$L = (m/2)\dot{X}^2 - \tfrac{1}{2}kX^2 + mgX$$

The generalized coordinate is obviously X, so that the appropriate Lagrange equation is

$$\frac{d}{dt}\left(\frac{\partial L}{\partial \dot{X}}\right) - \frac{\partial L}{\partial X} = Q_x$$

where

$$\frac{d}{dt}\left(\frac{\partial L}{\partial \dot{X}}\right) = m\ddot{X}$$

$$\frac{\partial L}{\partial X} = -kX + mg$$

$$Q_x = f_a(t)$$

The resulting differential equation is the familiar one

$$m\ddot{X} + kX - mg = f_a(t)$$

Figure 4.7 shows an alternate coordinate choice where $X = l + x$.

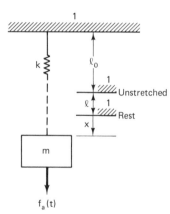

Figure 4.7 Vibrating Mass—No Damping

The incremental variable x in this coordinate system is measured from the static (rest) level of the mass rather than from the spring's unstretched level. Since l is a constant given by the static equilibrium equation

$$kl = mg$$

the second derivatives of X and x are related by

$$\ddot{X} = \ddot{x}$$

When X and $\ddot{X}$ are replaced by $l + x$ and $\ddot{x}$ in the differential equation, it becomes

$$m\ddot{x} + kx + kl - mg = f_a(t)$$

By using the static-equilibrium equation to cancel the terms $kl - mg$, the differential equation for the incremental variable is

$$m\ddot{x} + kx = f_a(t)$$

This differential equation is somewhat easier to solve than the original one. More importantly, it shows that if the problem had been formulated using the generalized coordinate x rather than X, it would not have been necessary to include the potential of the dead weight in the formulation of V.

To illustrate that this is not always the case, consider the pendulum of Figure 4.8. A linear torsional spring (not shown) joins the arm and ground at O_2. If this spring yields no torque when $\theta = 0$, then

$$V = \tfrac{1}{2}k\theta^2 - mgL \sin \theta$$

Assuming the arm is massless,

$$T = \tfrac{1}{2}m(L\dot{\theta})^2 + \tfrac{1}{2}I_{g_2}\dot{\theta}^2$$
$$= \tfrac{1}{2}(I_{g_2} + mL^2)\dot{\theta}^2$$
$$= \tfrac{1}{2}I_{O_2}\dot{\theta}^2$$

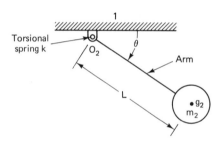

Figure 4.8   Pendulum with Torsional Spring

The applicable Lagrange equation is

$$\frac{d}{dt}\left(\frac{\partial L}{\partial \dot{\theta}}\right) - \frac{\partial L}{\partial \theta} = Q_\theta = 0$$

where

$$L = \tfrac{1}{2}I_{O_2}\dot{\theta}^2 - \tfrac{1}{2}k\theta^2 + mgL \sin \theta$$

The differential equation of motion is

$$I_{O_2}\ddot{\theta} + k\theta - mgL \cos \theta = 0$$

In words, this equation states that the spring torque and the rotational inertia oppose the moment of the gravity force.

The static equilibrium position, θ_o, is found by setting $\ddot{\theta} = 0$, which yields

$$k\theta_o - mgL \cos \theta_o = 0$$

Splitting θ into static (θ_o) and incremental dynamic (φ) parts, so that $\theta = \theta_o + \varphi$, changes the differential into the form

$$I_{O2}\ddot{\varphi} + k\varphi + k\theta_o - mgL \cos (\theta_o + \varphi) = 0$$

In this case, the introduction of an incremental variable does not lead to a simpler differential equation because the moment of the gravity force is nonlinear.

As a rule, incremental variables will not simplify the equation of motion when they contain nonlinear terms.

Example 2: Vibrating Mass with Damping

Figure 4.9(a) shows the previous spring-mass system with a linear dashpot b added external to the system. Figure 4.8(b) shows the force exerted on the mass by the dashpot opposing the motion. This force is treated as an external force, so that the previous equation of motion is now amended to read

$$m\ddot{x} + kx = f_a(t) - b\dot{x}$$

or, rearranging,

$$m\ddot{x} + b\dot{x} + kx = f_a(t)$$

which is the well-known equation for a damped mass–spring system.

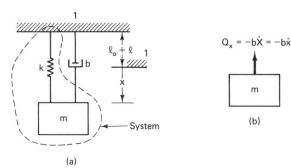

(a)

(b)

$$Q_x = -b\dot{X} = -b\dot{x}$$

Figure 4.9   Vibrating Mass—with Damping

Unfortunately, most dashpots rely upon viscous dissipation, with the result that they are better represented by the "square" law.

$$Q_X = -b\dot{x}|\dot{x}|$$

The absolute-value symbol is used so that the direction of Q_X will automatically change as the sign of x changes. The equation of motion for this system would now take the nonlinear form

$$mx + b\dot{x}|\dot{x}| + kx = f_a(t)$$

Solutions of these equations yield the velocity $\dot{x}(t)$ and displacement $x(t)$, from which the transmitted forces kx and $b|\dot{x}|\dot{x}$ can then be determined. Notice that the kinematics result from the formulation of a dynamic equation, after which the forces are found.

Example 3: Two-Mass System with Damping

Figure 4.10 shows two masses coupled together by a linear spring and a slow-moving (linear) dashpot. The system boundary has been drawn to exclude the dashpots and the mass-supporting lubrication layers so that they are external to the system. Although the masses are supported on fluid films, it will be assumed that they move slowly enough so that these films can also be treated as linear dashpots. Incremental coordinates x_1

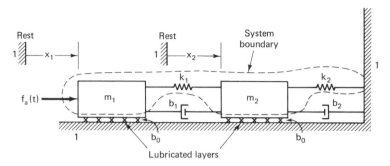

Figure 4.10   Two-Mass System with Damping

and x_2 are measured from the rest configuration that would be attained when $f_a(t) = 0$. The total kinetic energy of the system is

$$T = \tfrac{1}{2}m_1(\dot{x}_1)^2 + \tfrac{1}{2}m_2(\dot{x}_2)^2$$

The total potential energy is

$$V = \tfrac{1}{2}k_1(x_1 - x_2)^2 + \tfrac{1}{2}k_2x_2^2$$

Since we can express the dissipative effects in terms of $\dot{x}_1$ and $\dot{x}_2$ the generalized coordinates are x_1 and x_2. The Lagrange equations corresponding to these coordinates are

$$\frac{d}{dt}\left(\frac{\partial L}{\partial \dot{x}_1}\right) - \frac{\partial L}{\partial x_1} = Q_{x1}$$

$$\frac{d}{dt}\left(\frac{\partial L}{\partial \dot{x}_2}\right) - \frac{\partial L}{\partial x_2} = Q_{x2}$$

where

$$L = \tfrac{1}{2}m_1\dot{x}_1^2 + \tfrac{1}{2}m_2\dot{x}_2^2 - \tfrac{1}{2}k_1(x_1 - x_2)^2 - \tfrac{1}{2}k_2x_2^2$$

Performing the indicated operations on L yields two equations of motion, i.e.,

$$m_1\ddot{x}_1 + k_1(x_1 - x_2) = Q_{x1}$$

$$m_2\ddot{x}_2 + k_1(x_2 - x_1) + k_2x_2 = Q_{x2}$$

Figure 4.11 shows mass m_1 and the external forces that are applied to it. The dashpot force $b_1\,(\dot{x}_2 - \dot{x}_1)$ is shown acting to the right based on the assumption that $\dot{x}_2 > \dot{x}_1$, which would cause b_1 to stretch. If $\dot{x}_2 < \dot{x}_1$, this term will be negative, which corresponds to compressing b_1. The generalized force acting on m_1 affecting displacement x is

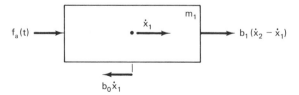

Figure 4.11   Forces Acting on m_1

$$Q_{x1} = f_a(t) + b_1(\dot{x}_2 - \dot{x}_1) - b_o(\dot{x}_1)$$

Figure 4.12 shows a similar force diagram drawn for mass m_2. The generalized force corresponding to the generalized coordinate x is

$$Q_{x2} = -b_1(\dot{x}_2 - \dot{x}_1) - (b_o + b_2)\dot{x}_2$$

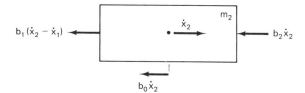

Figure 4.12   Forces Acting on m_2

With these generalized forces, the equations of motion become

$$m_1\ddot{x}_1 + b_1(\dot{x}_1 - \dot{x}_2) + b_o\dot{x}_1 + k_1(x_1 - x_2) = f_a(t)$$

$$m_2\ddot{x}_2 + b_1(\dot{x}_2 - \dot{x}_1) + (b_o + b_2)\dot{x}_2 + k_2x_2 = 0$$

The simultaneous solution of these linear coupled differential equations yields $x_1(t)$, $x_2(t)$, $\dot{x}_1(t)$, and $\dot{x}_2(t)$ from which the forces transmitted by the springs and dashpots can be determined.

Example 4: Traveling Inverted Pendulum

Figure 4.13 shows an inverted pendulum mounted on a cart of mass m_1. The massless arm of the pendulum is pinned at point (1) by a frictionless pin and restrained by a torsional spring k, which is unloaded when $\theta = 0$. The cart is propelled by applied force $f_a(t)$.

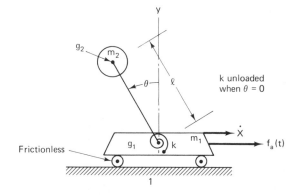

Figure 4.13   Traveling Inverted Pendulum

The velocity of the center of mass of m_2 is given by

$$\dot{\mathbf{V}}_{g2} = \dot{\mathbf{X}} + \dot{\boldsymbol{\theta}} \times \mathbf{l}$$

$$= (\dot{X} - l\dot{\theta} \cos \theta)\mathbf{i} - (l\dot{\theta} \sin \theta)\mathbf{j}$$

The kinetic energy for the two-mass system is

$$T = \tfrac{1}{2}m_2[(\dot{X} - l\dot{\theta} \cos \theta)^2 + (l\dot{\theta} \sin \theta)^2] + \tfrac{1}{2}(I_{zz})_2\dot{\theta}^2 + \tfrac{1}{2}m_1\dot{X}^2$$

where $(I_{zz})_2$ is the moment of inertia of mass m_2 about its own g.

In this example, it will be convenient to measure the angular displacement θ from the zero energy level of the torsional spring, which is not the static-equilibrium displacement. For that reason, the dead-weight potential *must* be included in the total potential, which is

$$V = \tfrac{1}{2}k\theta^2 + m_2 gl \cos \theta$$

The Lagrangian is then

$$L = T - V = \tfrac{1}{2}m_1 \dot{X}^2 + \tfrac{1}{2}m_2[\dot{X} - l\dot{\theta}\cos\theta)^2 + (l\dot{\theta}\sin\theta)^2]$$
$$+ \tfrac{1}{2}(I_{zz})_2 \dot{\theta}^2 - \tfrac{1}{2}k\theta^2 - m_2 gl \cos \theta$$

The coordinates X and θ are independent of each other and taken to be generalized coordinates. The corresponding Lagrange equations are

$$\frac{d}{dt}\left(\frac{\partial L}{\partial \dot{X}}\right) - \frac{\partial L}{\partial X} = Q_X$$

$$\frac{d}{dt}\left(\frac{\partial L}{\partial \dot{\theta}}\right) - \frac{\partial L}{\partial \theta} = Q_\theta$$

In the first equation, $Q_X = f_a(t)$. Performing the indicated operations yields

$$(m_1 + m_2)\ddot{X} - m_2 l\ddot{\theta}\cos\theta + m_2 l\dot{\theta}^2 \sin\theta = f_a(t)$$

The first term of this differential equation is the combined translational inertia that the two masses would have if m_2 did not rotate. The second and third terms are the horizontal components of the tangential and centrifugal inertias, respectively, of the rotating pendulum.

In the second Lagrange equation, $Q_\theta = 0$. The evaluation of the left-hand terms yields

$$[(I_{zz})_2 + m_2 l^2]\ddot{\theta} - (m_2 l \cos\theta)\ddot{X} + k\theta - m_2 gl \sin\theta = 0$$

In the first term of this equation, we find the expected parallel-axis shift term. The second term is the inertial moment of m_2 about g that would result if no rotation of m_2 occurred. The last two terms are the couple of the spring and the moment of the dead weight about g_1.

Solving the first equation for $\ddot{X}$ and inserting the result into the second equation yields a nonlinear second-order differential equation for $\theta(t)$. The solution of this differential equation can then be used to find $X(t)$.

Example 5: Rotating Pendulum

All of the previous examples have involved planar motion. Figure 4.14 shows a pendulum suspended midway on a rotating crossarm. The motion of mass m is constrained to a spherical surface like that of the earth, suggesting what may be a convenient set of generalized coordinates for this problem. Suppose that we use longitude and latitude lines as orthogonal coordinate directions along which the velocity components are measured. Then the velocity component along the longitude would be given by $l\dot{\theta}$, and the velocity component along the latitudes would be $l\Omega \sin \theta$, where Ω is the crossarm rotational speed, a function of time. With these, the kinetic energy of the pendulum mass is

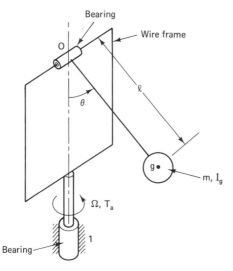

Figure 4.14 Rotating Pendulum

$$T = \tfrac{1}{2}m[(l\Omega \sin \theta)^2 + (l\dot\theta)^2] + \tfrac{1}{2}I_g(\dot\theta^2 + \Omega^2)$$

If the zero level of potential energy is taken to be the rest level, the potential energy can be written as

$$V = mgl(1 - \cos \theta)$$

By combining T and V, the Lagrangian is

$$L = \tfrac{1}{2}m[(l\dot\theta)^2 + (l\Omega \sin \theta)^2] + \tfrac{1}{2}I_g(\dot\theta^2 + \Omega^2) - mgl(1 - \cos \theta)$$

This choice of coordinates has led to a relatively simple formulation of the Lagrangian. There are other possible choices of independent coordinates, all of which will lead to the same motion. Among these alternatives, one usually proves to be easier to work with than the rest.

One of the appropriate Lagrange equations is

$$\frac{d}{dt}\left(\frac{\partial L}{\partial \dot\theta}\right) - \frac{\partial L}{\partial \theta} = Q_\theta$$

where $Q_\theta = 0$. Inserting L into this equation yields

$$(I_g + ml^2)\ddot\theta - ml^2\Omega^2(\sin \theta \cos \theta) + mgl \sin \theta = 0$$

Figure 4.15 shows the external forces that do work on the system. (The tensile force along l does no work on the system.) The last two terms of the differential equation are the moments of these forces about the hinge point O. To start, the motion Ω must change with time. However, if Ω is then leveled off to some constant value, all dynamic effects should disappear. In this equation, θ would then be a constant and $\dot\theta = \ddot\theta = 0$, leaving only the last two terms. These two terms automatically satisfy static equilibrium of external force moments about O.

All this illustrates that the dynamic equations should reduce to equations of static equilibrium when the terms associated with dynamic effects are removed. One should always test the equations of motion to see if they satisfy static equilibrium when the

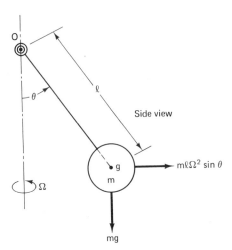

Side view

$m\ell\Omega^2 \sin\theta$

Figure 4.15 External Forces which Do
Work on the System

derivatives and the forces and moments that cause them are deleted from the equations
of motion.

Since the applied torque T_a does not appear in this equation, there must be another equation of motion in which it does. The Lagrange equation for the other generalized coordinate provides this relationship. Because the Lagrangian contains only the
derivative (Ω) of the other generalized coordinate, the torque required to obtain the
motion is given by the Lagrange equation

$$\frac{d}{dt}\left(\frac{\partial L}{\partial \Omega}\right) = Q_\Omega = T_a$$

which becomes

$$\frac{d}{dt}\{[I_g + (l \sin\theta)^2 m]\Omega\} = T_a$$

This equation of motion can also be written in the form

$$[I_g + (l \sin\theta)^2 m]\Omega = \int_0^t T_a \, dt$$

where it is assumed that Ω is initially zero. During startup, T_a is greater than zero. If,
after startup, T_a is reduced to zero, the angular momentum $[I_g + (l \sin\theta)^2 m]\Omega$ will become constant.

Example 6: Rolling Cylinder without Slip

Figure 4.16 shows the now familiar rolling cylinder analyzed in the previous chapter.
Contrary to the previous examples where friction extracted work from the system,
rolling friction that causes a reversible elastic deformation of the contact areas does not
result in the dissipation of energy from the system. The normal and friction forces are
then workless constraints, so that they will not appear in Q_x or Q_y.

If the cylinder rolls without slip, the velocity of the center of mass is
$\dot{\mathbf{R}}_g = \dot{\boldsymbol{\theta}} \times \mathbf{D}$. The speed of the center of mass is the scalar $\dot{\theta}D$. The total kinetic energy of the cylinder is

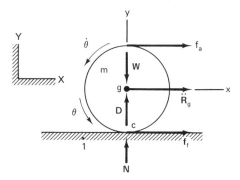

Figure 4.16 Rolling Cylinder

$$T = \tfrac{1}{2}m\dot{R}_g^2 + \tfrac{1}{2}I_{zz}\dot{\theta}^2 = \tfrac{1}{2}(mD^2 + I_{zz})\dot{\theta}^2$$

Note the appearance of the parallel-axis theorem in this formulation. This is not the only possible formulation, however, since an equally valid alternative form is

$$T = \frac{1}{2}(mD^2 + I_{zz})\left(\frac{\dot{R}_g}{D}\right)^2$$

in which the state variable is $\dot{R}_g$ instead of $\dot{\theta}$.

Since the center of mass does not change elevation, the potential energy is constant, say, zero. The Lagrangian is then

$$L = \tfrac{1}{2}(mD^2 + I_{zz})\dot{\theta}^2 = \frac{1}{2}\left(\frac{mD^2 + I_{zz}}{D^2}\right)\dot{X}^2$$

where $\dot{R}_g$ has been replaced by $\dot{X}$, so that X measures the displacement of the center of mass. The appropriate Lagrange equation when X is the generalized coordinate is

$$\frac{d}{dt}\left(\frac{\partial L}{\partial \dot{X}}\right) - \frac{\partial L}{\partial X} = Q_X$$

The equation of motion is, therefore,

$$\left(\frac{mD^2 + I_{zz}}{D^2}\right)\ddot{X} = Q_X$$

The force Q_X is the force that gives rise to the displacement dX of the center of mass, which is located by the generalized coordinate X shown in Figure 4.16. The displacement of the top of the cylinder is twice the displacement of g, so that the incremental energy input to the cylinder by f_a is

$$dW = 2f_a\, dX$$

and therefore $Q_X = 2f_a$.

The equation of motion is now

$$(mD^2 + I_{zz})\ddot{X} = 2f_a D^2$$

This equation can be made to coincide with the formulation of the last chapter if we note that

$$\ddot{X} = -\ddot{\theta}D$$

Before summarizing what we have discoverd from this analysis, we will consider what might have been done with the alternate kinetic-energy formulation $T = \frac{1}{2}(mD^2 + I_{zz})\dot{\theta}^2$. The generalized coordinate is now θ. The appropriate Lagrange equation is now

$$\frac{d}{dt}\left(\frac{\partial L}{\partial \dot{\theta}}\right) - \frac{\partial L}{\partial \theta} = Q_\theta$$

which reduces to

$$(mD^2 + I_{zz})\ddot{\theta} = Q_\theta$$

The left-hand side of this equation is a couple since $\dot{\theta}\mathbf{k}$ is a free vector. The right-hand side is also a couple as noted before. Figure 4.17(a) shows $\mathbf{f}_a$ and $-\mathbf{f}_a$ added to the original forces at the constraining surface c to form an "equivalent force" system. Figure 4.17(b) shows this same system reduced to a couple $2\mathbf{D} \times \mathbf{f}_a$ and a workless force $\mathbf{f}_a$ applied at the constraining surface c. Since this couple is contrary to the positive rotational sense, $Q_\theta = -2Df_a$, so that the Lagrange equation becomes

$$(mD^2 + I_{zz})\ddot{\theta} + 2Df_a = 0$$

as before because $\ddot{X} = -\ddot{\theta}D$.

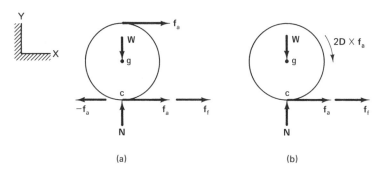

Figure 4.17 Equivalent System—No Slip

Some will correctly argue that $2Df_a$ is the moment of f_a about c. However, there was no basis for preferring this couple over the couple Df_a or any other couple before the equivalent system of Figure 4.17(b) was created. Any other equivalent system would have left f_a applied to a moving point just as it was in the original problem and nothing would have been gained.

Example 7: Rolling Cylinder with Slip

In the no-slip case just analyzed, $\dot{X}$ and $\dot{\theta}$ are not independent. We demonstrated that the equation of motion could be obtained by writing the kinetic energy in terms of one of these two variables and then applying the appropriate Lagrange equation. When there is slip, $\dot{X}$ and $\dot{\theta}$ cannot be related to each other by a kinematic expression such as $\dot{X} = \dot{\boldsymbol{\theta}} \times \mathbf{D}$. In that case, the relationship between $\dot{X}$ and $\dot{\theta}$ is determined by the dynamics of the system and not the kinematics alone, as in the previous problem. That means that we have to solve the dynamics problem *before* we can determine an expression relating X and θ.

The formulation of the kinetic energy with slip must remain in its primitive form:

$$T = \tfrac{1}{2}mV_g^2 + \tfrac{1}{2}I_{zz}\dot{\theta}^2$$

or

$$T = \tfrac{1}{2}m\dot{X}^2 + \tfrac{1}{2}I_{zz}\dot{\theta}^2$$

The generalized coordinates are now X and θ. Two Lagrange equations now apply. They are

$$\frac{d}{dt}\left(\frac{\partial L}{\partial \dot{\theta}}\right) - \frac{\partial L}{\partial \theta} = Q_\theta$$

and

$$\frac{d}{dt}\left(\frac{\partial L}{\partial \dot{X}}\right) - \frac{\partial L}{\partial X} = Q_X$$

These equations reduce to

$$I_{zz}\ddot{\theta} = Q_\theta$$
$$m\ddot{X} = Q_X$$

The last equation calls for the force acting through a distance dX. Figure 4.18(b) shows the equivalent system for Figure 4.18(a) with $\mathbf{f}_a$ and $\mathbf{f}_f$ acting at g and couples $\mathbf{D} \times \mathbf{f}_f$ and $\mathbf{D} \times \mathbf{f}_a$.

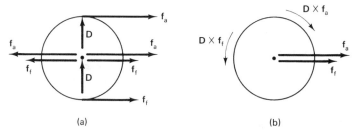

(a)   (b)

Figure 4.18   Equivalent System—Slip

Then

$$Q_X = f_a + f_f = f_a + \mu_d W$$
$$Q_\theta = D(f_f - f_a) = D(\mu_d W - f_a)$$

so that the equations of motion become

$$f_a + \mu_d W - m\ddot{X} = 0$$
$$-f_a D + \mu_d W D - I_{zz}\ddot{\theta} = 0$$

Example 8: Spinning Top

In all of the previous examples, the kinetic and potential energies were relatively easy to formulate. The spinning top requires considerable care in the formulation of its kinetic and potential energies, as we shall see.

In Figure 4.19, X, Y, and Z are fixed coordinates, whereas the x, y, z coordinates are centered at g and move with the top so that z is always along the axis of symmetry, x is always parallel to the X–Y plane and y is always orthogonal to x and z.

The top spins around z with angular velocity $\dot{\phi}$, precesses about Z with $\dot{\psi}$, and rotates about x with $\dot{\theta}$. The angles corresponding to these velocities are called the Euler angles. They are chosen to describe the top's motion because they are easily observed and measured. Unfortunately, they are not mutually orthogonal and, therefore, cannot be used directly to formulate the kinetic energy of the top.

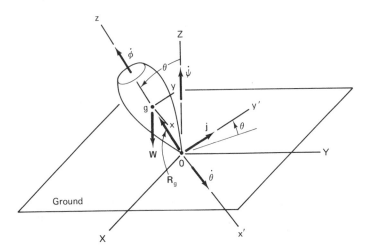

Figure 4.19   Spinning Top

There are two orthogonal coordinate systems shown in Figure 4.19, the stationary XYZ system and the moving xyz system. Since ϕ and θ are already directed along z and x, respectively, the moving coordinate system will be selected as a basis for constructing the angular velocity of the top. The xyz coordinates are also principal coordinates; hence, there are no products of inertia relative to this coordinate system. Furthermore, the moments of inertia of the top are constant when viewed from this system, so that their time derivatives are zero. As a consequence of this choice, the kinetic energy formulated in this system will be relatively simple and its time derivative will involve only derivatives of the angular-velocity components along the coordinates.

From Figure 4.19 the components of the angular velocity of the top along the x, y, and z coordinates are

$$\omega_x = \dot{\theta}$$

$$\omega_y = \dot{\psi} \sin \theta$$

$$\omega_z = \dot{\phi} + \dot{\psi} \cos \theta$$

The moments of inertia about x, y, and z (relative to the center of mass) are I_{xx}, I_{yy}, and I_{zz}, respectively. The rotational kinetic energy relative to the center of mass is, therefore,

$$\tfrac{1}{2}[I_{xx}\omega_x^2 + I_{yy}\omega_y^2 + I_{zz}\omega_z^2]$$

where $I_{xx} = I_{yy}$ due to symmetry. This portion of the total energy can also be written in terms of the generalized nonorthogonal coordinates θ, ϕ, and ψ as

$$\tfrac{1}{2}\{I_{xx}[\dot\theta^2 + (\dot\psi \sin \theta)^2] + I_{zz}(\dot\phi + \dot\psi \cos \theta)^2\}$$

The angular velocity of the xyz coordinate system (fixed in the top) is given by

$$\mathbf{\Omega} = (\dot\theta)\mathbf{i} + (\dot\psi \sin \theta)\mathbf{j}$$

and its center of mass is located by

$$\mathbf{R}_g = R_g \mathbf{k}$$

so that the velocity the center of mass g is

$$\dot{\mathbf{R}}_g = \mathbf{\Omega} \times \mathbf{R}_g = (R_g \dot\psi \sin)\mathbf{i} - (R_g \dot\theta)\mathbf{j}$$

The kinetic energy of the center of mass is, therefore,

$$\tfrac{1}{2}m\dot{R}_g^2 = \tfrac{1}{2}m[(R_g \dot\psi \sin \theta)^2 + (R_g \dot\theta)^2]$$

The total kinetic energy is then

$$T = \tfrac{1}{2}[(I_{xx} + mR_g^2)\dot\theta^2 + (I_{xx} + mR_g^2)(\dot\psi \sin \theta)^2 + I_{zz}(\dot\phi + \dot\psi \cos \theta)^2]$$

From this expression, we see that the total kinetic energy can be expressed in terms of the moments of inertia relative to the primed coordinates x', y', and z with its origin at 0 having angular velocity of $\mathbf{\Omega} = (\dot\theta)\mathbf{i} + (\dot\psi \sin \theta)\mathbf{j}$. Designating these as

$$I'_{xx} = (I_{xx} + mR_g^2)$$

$$I'_{zz} = I_{zz}$$

the total kinetic energy is simply

$$T = \tfrac{1}{2}\{I'_{xx}[\dot\theta^2 + (\dot\psi \sin \theta)^2] + I'_{zz}(\dot\phi + \dot\psi \cos \theta)^2\}$$

The potential energy of the top is

$$V = m_g R_g \cos \theta$$

so that the Lagrangian becomes

$$L = \tfrac{1}{2}\{I'_{xx}[\dot\theta^2 + (\dot\psi \sin \theta)^2] + I'_{zz}(\dot\phi + \dot\psi \cos \theta)^2\} - mgR_g \cos \theta$$

The Lagrange equations for the generalized coordinates θ, ϕ, and ψ are

$$\frac{d}{dt}\left(\frac{\partial L}{\partial \dot\theta}\right) - \frac{\partial L}{\partial \theta} = Q_\theta$$

$$\frac{d}{dt}\left(\frac{\partial L}{\partial \dot\phi}\right) - \frac{\partial L}{\partial \phi} = Q_\phi$$

$$\frac{d}{dt}\left(\frac{\partial L}{\partial \dot\psi}\right) - \frac{\partial L}{\partial \psi} = Q_\psi$$

In these equations, Q_θ, Q_ϕ, and Q_ψ are the torques (couples) applied to the top along the θ, ϕ, and ψ axes, respectively. For an unpowered (coasting) top on a frictionless support at "o" they would all be zero. The Lagrange equations would then reduce to

$$I'_{xx}\ddot{\theta} - I'_{xx}\dot{\psi}^2 \sin\theta\cos\theta + I'_{zz}(\dot{\phi} + \dot{\psi}\cos\theta)(\dot{\psi}\sin\theta) - mgR_g\sin\theta = 0$$

$$\frac{d}{dt}[\dot{\phi} + \dot{\psi}\cos\theta] = 0$$

$$\frac{d}{dt}[I'_{xx}(\dot{\psi}\sin^2\theta) + I'_{zz}(\dot{\phi} + \dot{\psi}\cos\theta)\cos\theta] = 0$$

The last two equations are particularly convenient. If Q_ϕ and Q_ψ are given functions of time, these two equations can be integrated once with respect to time, reducing them to first-order equations.

4.7 RECAPITULATION

Several general features of the Lagrange equations are illustrated by these examples. The kinetic and potential energies are relatively easy to formulate. In multidimensional problems, the most common errors result from either not accounting for all the energy or from using nonorthogonal velocity components to formulate T.

Probably the most useful feature of the method is the flexibility in the choice of generalized coordinates. There are usually several possible sets of generalized coordinates that could be used, some easier to implement than others.

Once it has been verified that the coordinates used to formulate T and V are kinematically independent, the form and number of Lagrange equations are immediately apparent. Note that the formulation of the kinetic energy requires a kinematic description of the system only up to the level of velocities. It is not necessary to formulate the accelerations of the body, since that is done automatically by the differentiations called for in the Lagrange equations.

The generalized forces are usually the most difficult to formulate. These must be related to the generalized coordinates through the work concept.

The end results of the Lagrange equations are always linear-momentum and angular-momentum equations. After obtaining the equations, one should go through the equations term by term, identifying the source of each term, be it a force, moment, or couple. Embarrassing errors can often be eliminated by a careful inspection of the physical meaning of each term in the resulting equations. One should also verify that the dynamic equations reduce to the static-equilibrium equations for the system when the dynamic terms are deleted. In linear cases where incremental variables are used, both sides of the equations should become identically zero.

In multi-element systems, the interconnecting forces and moments internal to the system are self-canceling if they are frictionless. The only external forces and moments that contribute to the generalized forces are those that add to or subtract energy from the systems.

4.8 LINEARIZATION OF THE DIFFERENTIAL EQUATIONS OF MOTION

In each of the examples examined, the end results were always ordinary differential equations for the generalized coordinates. These equations are "driven" by known forces and couples. Their solutions yield the displacements, velocities, and accelerations of each element of the system.

The solution of these differential equations by analytical means is not always easy or straightforward. Today, with the advent of high-speed computers that can rapidly execute various integration schemes, their numerical solution becomes relatively routine. Once the kinematic quantities are established, the interconnecting forces can be determined from an analysis of the element free-body diagrams using the d'Alembert technique. If the mechanism is planar, the graphic techniques of the previous chapter can be implemented.

Before launching headlong into a lengthy computer solution of the differential equations of motion, it is often instructive to linearize the equations about an operating point, if one exists.

In the case of Example 4, where the equations of motion for the traveling inverted pendulum were

$$(m_1 + m_2)\ddot{X} - m_2 l\ddot{\theta} \cos \theta + m_2 l\dot{\theta}^2 \sin \theta = f_a(t)$$

$$[(I_{zz})_2 + m_2 l^2]\ddot{\theta} - m_2 l \cos \theta \ddot{X} + k\theta - m_2 gl \sin \theta = 0$$

a stiff torsional spring k coupled with a small mass m_2, a large mass m_1, and a small force $f_a(t)$ would probably result in small angular deflections θ. Under these circumstances, we could make the straightforward approximations that $\sin \theta \sim \theta$ and $\cos \theta \sim 1$. If we also take the amplitude of θ to be $\theta_o \ll 1$, then $\ddot{\theta}$ would be proportional to θ_o and $\dot{\theta}^2$ would be proportional to θ_o^2. As a result, in the first equation, the term $m_2 l\dot{\theta}^2 \sin \theta$ is small compared to the term $m_2 l\ddot{\theta} \cos \theta$ and, therefore, can be omitted.

With these approximations, the linearized equations of motion become

$$(m_1 + m_2)\ddot{X} - (m_2 l)\ddot{\theta} = f_a(t)$$

$$[(I_{zz})_2 + m_2 l^2]\ddot{\theta} - (m_2 l)\ddot{X} + (k - m_2 gl)\theta = 0$$

Now combining the two equations to eliminate $\ddot{X}$ yields the differential equation

$$\left[(I_{zz})_2 + m_2 l^2 - \frac{(m_2 l)^2}{m_1 + m_2}\right]\ddot{\theta} + (k - m_2 gl)\theta = \left(\frac{m_2 l}{m_1 + m_2}\right)f_a(t)$$

which is linear and has constant coefficients. The solution of this equation with its initial conditions will reveal the approximate behavior of the system when θ is small. Since the solution is relatively easy to perform, a great deal can be learned about the response of the system from a modest investment of effort.

This linearized solution can also be used to check the results of computer solutions. If the computer program is an accurate simulation, it should approach the approximate analytical solution for the cases where θ is small.

The linearization of the rotating-pendulum equation in Example 5 for small angular displacements θ when $ml^2 \gg I_g$ reduces it to

$$\ddot{\theta} + (g/l - \Omega^2)\theta = 0$$

When Ω is constant, the exact static-equilibrium position is given by $\cos \theta = g/l/\Omega^2$, which means that there are no static-equilibrium positions for $g/l > \Omega^2$. The solution of the linearized differential equation for $g/l > \Omega^2$ is

$$\theta = c_1 \sin (g/l - \Omega^2)^{1/2}t + c_2 \cos (g/l - \Omega^2)^{1/2}t$$

where c_1 and c_2 are arbitrary constants that depend on the initial conditions.

This is an oscillatory (stable) motion about $\theta = 0$, with a frequency $(g/l - \Omega^2)^{1/2}$. It is a reasonably accurate representation of the response provided the amplitude of θ remains small, say, less than $10°$. We might guess that larger displacements are probably also oscillatory as long as $g/l > \Omega^2$.

When $\Omega^2 > g/l$, the differential-equation solution is

$$\theta = c_1 \sinh (\Omega^2 - g/l)^{1/2}t + c_2 \cosh (\Omega^2 - g/l)^{1/2}t$$

Since the hyperbolic functions grow without bound, initially, small disturbances in the displacement cannot remain small. To put it another way, any small perturbation of θ from $\theta = 0$ yields an unstable condition that drives θ to displacements that cannot be considered small. Under these circumstances, the nonoscillatory motion of the pendulum sends it on a search for the static-equilibrium position when Ω is constant. The path of this search is not predicted, even approximately, by the previously linearized solution.

In both these examples, linearization has reduced the differential equations to easily solvable forms. The solutions, restricted as they may be, can provide considerable insight into the behavior of the system. In many applications, the decision to proceed with a more detailed analysis can be based on the results of linearized solutions. If certain undesirable performance characteristics of the system have already been revealed by the linearized results, it makes little sense to continue any further analyses.

REFERENCES

1. Huang, T. C. *Engineering Mechanics*. Reading, MA: Addison-Wesley, 1967.
2. Spiegel, M. R. *Theoretical Mechanics*. New York: Schaum, 1967.
3. Greenwood, D. T. *Principles of Dynamics*, 2nd Ed. Englewood Cliffs, NJ: Prentice Hall, 1987.

PROBLEMS

All of these problems are to be done using the Lagrange equations as the starting point, not Newton's laws.

4.1. The system in the diagram shows a mass–spring system forced sinusoidally. Obtain the equations of motion. The springs are linear and the supports are frictionless.

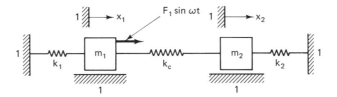

4.2. Obtain the equations of motion for the system that contains a linear dashpot B, as shown in the diagram. Masses m_1 and m_2 are on frictionless rollers. The spring is linear.

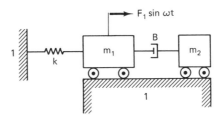

4.3. The diagram shows a torsional system that is excited by a sinusoidal torque. Derive the equations of motion. The torsional springs are linear. J is the polar moment of inertia.

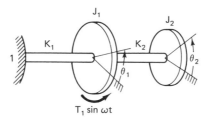

4.4. Determine the equation of motion of the system shown in the diagram. Determine the natural frequency of this system. The pulley's radius of gyration is $R_G = 4$ in.

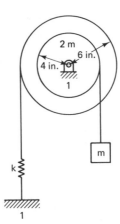

4.5. The mass–spring system shown in the diagram is driven by a sinusiodally varying force applied to the pin joining the springs at C.

  (a) Obtain the equations of motion. *Hint:* Place a mass at node C between the springs and set that mass equal to zero in the resulting equations of motion. Then eliminate x_C from the remaining equations.

  (b) Find the natural frequencies of the system.

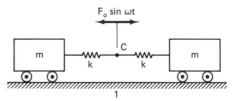

4.6. The system in the diagram is in its static-equilibrium configuration. Assuming that the cylinder (2) rolls without slipping, obtain the equation of motion when a clockwise couple $T_o \sin \omega t$ is applied to the cylinder. Also assume that T_o is small so that the angular displacement of the cylinder is small. What value of ω will cause this assumption to be incorrect?

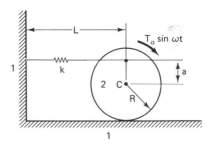

4.7. The small bead m_2 shown in the diagram is free to slide along the thin rigid rod of length L. Small mass m_1 is attached to the rod. The bead is constrained to slide on the frictionless horizontal surface (1) while it slides along the rod. Obtain the natural frequency of this system for small θ displacements.

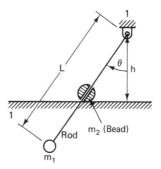

4.8. The cart m_1 is caused to oscillate by the attached pendulum, as shown in the diagram. The pendulum bob m_2 is much heavier than the arm, so that its center of mass g_2 can be assumed to be at the center of the bob. The cart wheels are frictionless. Find the equations of motion.

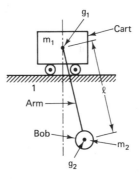

4.9. The cart of Problem 4.8 is replaced by a sliding block m_1 restrained by two equal linear springs k. The friction between the block and its guideway is nonlinear, i.e., it is proportional to the velocity squared. The proportionality constant is B. Obtain the equations of motion.

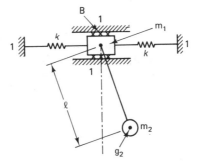

4.10. A semicircular cylinder rolls without slipping. Obtain the equations of motion. What is the natural frequency of the semicylinder for small displacements from rest?

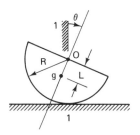

4.11. A disk (3) is attached to the end of the arm (2). Arm (1) is rigidly connected to arm (2) to form a right angle at the frictionless pivot. A linear spring k and a linear dashpot B (force is proportional to velocity) are connected, as shown in the diagram. An oscillating couple $T(t)$ is applied at the pivot. The diagram shows the system in its rest position. Assuming small angular displacements, determine the following:

  (a) The system's kinetic energy.
  (b) The system's potential energy.
  (c) Identify your generalized coordinates.

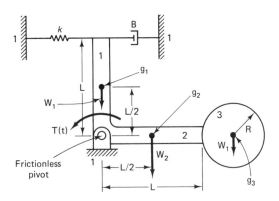

4.12. A sphere (2) is attached to the end of arm (1), as shown in the diagram. The arm is connected to the wall by a frictionless pivot. A torsional spring k and a linear dashpot B (torque is proportional to angular velocity) are connected to the arm as shown. An oscillating couple $T(t)$ is applied at the pivot. The diagram shows the system in its rest position. Assuming small angular displacements, determine the following:

  (a) The system's kinetic energy.
  (b) The system's potential energy.
  (c) Identify your generalized coordinates.
  (d) Identify the generalized forces.
  (e) Obtain the equations of motion.

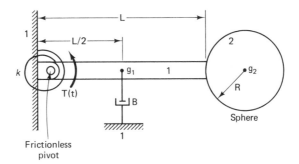

4.13. The diagram shows a common toy called a yo-yo. The disk-shaped mass m is wrapped with a string that reaches vertically to a person's hand h. By applying a oscillating vertical force to the string at h, the mass m is caused to oscillate up and down. Consider $F_a(t) = F_o + F_1 \sin \omega t$ to be the "cause" and the vertical velocity of h (attached to the hand) and the rotation of the disk to be the "effects". The string is massless and always taut (in tension) since $F_o > F_1$. It may be convenient to consider the weight of the disk to be an external force. Obtain the following:

(a) The kinetic energy.

(b) The potential energy.

(c) The Q's.

(d) The differential equations.

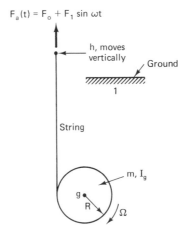

4.14. A diagram of a flyball governor is shown. The spherical masses m_b are rigidly attached to links (2) and pinned to links (3). Links (2) are pinned to the vertical shaft at a. Links (3) are pinned to the sliding collar m_c at b. The collar is joined to a, fixed on the shaft, by a spring k that is wound around the vertical shaft. The length of the spring when not loaded is l. Rotation of the massless links and shaft causes the ball masses to move outward. I_g is not negligible for m_b or m_c.

(a) Formulate the kinetic energy for the system.

(b) Formulate the potential energy for the system.

(c) Using Lagrange methods, obtain the differential equation of motion that contains the applied torque $T_a(t)$.

(d) Using Lagrange methods, obtain the equation for the static value of ψ.

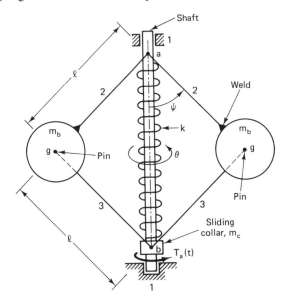

4.15. Find the equations of motion for the Scotch-yoke mechanism shown in the diagram. The spring is preloaded so that the spring is stretched even when $\theta = 0$. Linearize the differential equation of motion for small displacements from static equilibrium. Determine the natural frequency of these displacements.

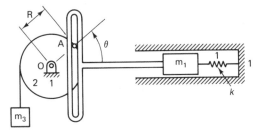

4.16. A disk (1) rotates in a horizontal plane with $\dot{\theta}$ and $\ddot{\theta}$ about point O, as shown in the diagram. Attached to the rim of the disk at a is a uniform bar. With no rotation, the coiled spring at a maintains $\phi = 0$. It supplies a restorting torque $= k\phi$. Torque $T_a(t)$ is applied to the disk at O.

(a) Set up the equation for the kinetic energy.

(b) Set up the equation for the potential energy.

(c) Using generalized coordinates θ and $\theta - \phi = \psi$, obtain the equations of motion.

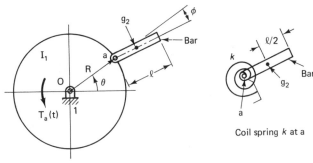

Plan view

4.17. The diagram shows a disk *m* supported at its perimeter by a massless arm. A pin at *C* allows the disk to rotate freely so that the system forms a pendulum with a loose disk. Oscillating torque $T_s(t)$ is applied to the arm at *O* to rotate the system in the vertical plane. The pin at *O* is also frictionless.

(a) Write the Lagrange equation whose driving torque is $T_s(t)$.

(b) Using the results of part (a), select the appropriate coordinates from those shown in the diagram and
   (i) formulate the kinetic energy of the system that could be used in the equation of part (a).
   (ii) formulate the potential energy of the system that could be used in the equation of part (a).

(c) Obtain the differential equation containing $T_s(t)$.

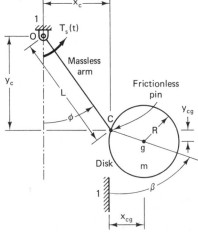

4.18. The diagram shows a carnival ride. It consists of a rotating arm (2) and a bucket (3) pinned to it at *C*. There is a torsional spring a *C* that is not shown. The spring is relaxed when a single passenger enters the bucket, shifting the center of mass of the passenger–bucket combination to g_3. The arm is then started and rotates at $\dot{\theta} =$ constant.

(a) Obtain the kinetic energy of the system.

(b) Obtain the potential energy of the system.

(c) Obtain the equation for the required torque T_o.

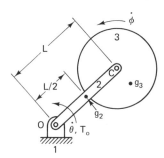

4.19. The system shown in the diagram is excited by the applied force $F_a(t)$. There is friction proportional to velocity between m_1 and its guideway and at the pin. The spring k is linear. Determine the equations of motion relative to the rest positions. Determine the stretch of the spring in the rest position.

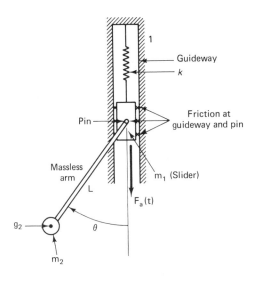

4.20. The cart m_1 and the spring/pendulum m_2 are propelled by an applied force $f_a(t)$ that acts at the tip of the pendulum, a distance L from the pin at O. The slotted link (L long) is massless. Mass m_2 has negligible I_{g2}. The tip of the spring (s) is a distance l from O when mass m_2 is attached to the spring k. Link (L) is vertical during assembly.
   (a) Obtain T for the system.
   (b) Obtain V for the system.
   (c) Obtain the Q's for the system.
   (d) Obtain the differential equation that results from

$$\frac{d}{dt}\left(\frac{\partial L}{\partial \dot{X}}\right) - \frac{\partial L}{\partial X} = Q_X$$

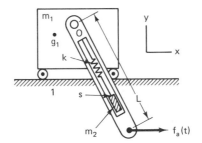

4.21. The diagram shows a cylinder m_2 that rolls without slipping at C on wedge m_3. The cylinder is restrained by spring k, whose unloaded length is X_o. The motion of m_2 causes m_3 to move horizontally on its frictionless air-bearing support. Obtain the equations of motion for this system.

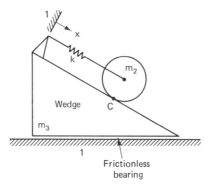

4.22. The disk shown in the diagram rotates about the vertical, driven by torque $T_a(t)$. Link (2) is pinned to the rim of the disk. A torsional spring k_ϕ (not shown) joins the link (2) to disk (1) at the pin. The spring is unloaded when $\phi = 0$. The pin is frictionless.
(a) Formulate the kinetic energy for the disk and link system.
(b) Formulate the potential energy for the system.
(c) For high-speed operation ($\dot{\psi}$ large) with $\ddot{\psi} = 0$ and ϕ small, find the natural frequency of link (2) motion.

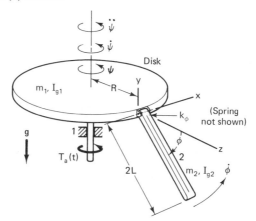

4.23. Consider a small bead of mass m that is free to slide without friction along a circular hoop under the action of a gravitational field oriented along the vertical axis shown in the diagram. The hoop is caused to rotate at constant angular velocity ω. Determine the equations of motion of the mass m relative to the spin axis using angular coordinate ϕ.

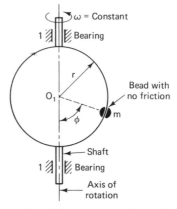

4.24. A disk of diameter D and radius R is pinned to the end of arm OA of length L, as shown in the diagram. The arm disk rotates about the pin through A. Obtain the equations of motion for this system.

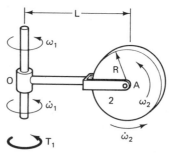

4.25. The diagram shows a pendulum m_2 mounted to ground with a frictionless pin at O_2. The fixed center of mass of m_2 is at g_2. A torsional spring connects m_2 to ground so that the stable rest position of the pendulum is vertical and underneath point O_2, as shown. A movable mass m_3 is supported within the pendulum by spring k_r. When unloaded, the length of the spring is r_o. Its center of mass is g_3. When the pendulum is set into motion, m_3 can move relative to the slot, without friction. Masses m_2 and m_3 are of comparable size.

(a) Write an expression for the total kinetic energy of the system when in motion.

(b) Write an expression for the total potential energy of the system when in motion.

(c) Using the appropriate Lagrange equations, obtain the equations of motion. Write the final form of these equations measured in terms of independent generalized coordinates.

Note: The line joining O_2, g_3, and g_2 is in the direction of g (gravity) at rest.

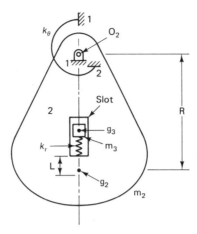

4.26. Cylinder (2) rolls in the slot without slipping at C, as shown in the diagram. It is restrained by spring k, whose unstretched length is r_o. The slot is in arm (1) that rotates about the fixed point O due to torque T_s.
  (a) Determine the kinetic energy of m_1.
  (b) Determine the kinetic energy of m_2.
  (c) Determine the potential energy of the system.
  (d) What are the generalized coordinates?
  (e) What are the generalized "forces"?

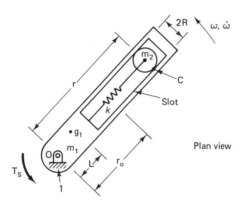

4.27. The diagram shows a portion of a disk that rotates about the vertical axis a–a. A block of mass m slides in the radial slot. Friction forces are exerted on the bottom of the block and on the side of the block because of the disk's acceleration. The disk is massless. Using a cylindrical coordinate system and Lagrange methods, determine the following:
  (a) Obtain the differential equation for the tangential motion of m.
  (b) Obtain the differential equation for the radial motion of m.

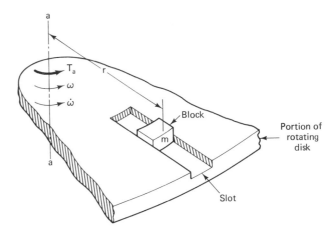

4.28. Lagrange equations may be used to derive the equations of an electrical network when the elements can be lumped into inductances (L), capacitances (C), and resistances (R). Using charge q as a generalized coordinate, develop the following:

   (a) The electrical analog of kinetic energy.

   (b) The electrical analog of a spring when there is no initial energy storage.

   (c) Modify part (b) to include the effect of initially stored energy.

4.29. For the *LCR* circuit shown:

   (a) Formulate the Lagrangian.

   (b) Formulate the generalized "force."

   (c) Using the appropriate Lagrange equation, derive the equation for the circuit. Include the dissipative effect of resistance R.

   (d) Indicate how $E(t)$ can be treated as a potential.

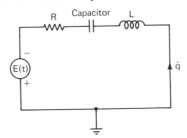

4.30. Using the methods of Lagrange obtain the equation for the network shown. Note that $\dot{q}_1$, $\dot{q}_2$, and $\dot{q}_3$ are not independent.

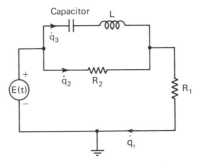

APPENDIX: Lagrange Equations

If we consider a conservative system, the sum of the kinetic and potential energies is constant and the differential of the sum is zero, i.e.,

$$d(T + V) = 0$$

where T is the total kinetic energy, and V is the total potential energy.

The potential energy V is usually expressed in terms of elastic displacements or generalized coordinates measured. Denoting the generalized coordinates as $q_i (i = 1, 2, 3, \ldots, n)$, then

$$V = V(q_1, q_2, q_3, \ldots, q_n)$$

The kinetic energy is a function of the generalized coordinates and velocities so that

$$T = T(q_1, q_2, q_3, \ldots, q_n, \dot{q}_1, \dot{q}_2, \dot{q}_3, \ldots, \dot{q}_n)$$

The differential of T is

$$dT = \sum_{i=1}^{n} \frac{\partial T}{\partial q_i} dq_i + \sum_{i=1}^{n} \frac{\partial T}{\partial \dot{q}_i} d\dot{q}_i$$

The term $\sum_{i=1}^{n} (\partial T / \partial \dot{q}) i \, d\dot{q}i$ can be eliminated from this expression if we consider the differential of the generalized expression for the kinematic energy, which is

$$T = \tfrac{1}{2} \sum_{i=1}^{n} \sum_{j=1}^{n} m_{ij} \dot{q}_i \dot{q}_j$$

in which m_{ij} is the function of the mass and generalized coordinates. Differentiating this formula with respect to $\dot{q}_i$ and finally summing from $i = 1$ to n yields

$$\sum_{i=1}^{n} \frac{\partial T}{\partial \dot{q}_i} \dot{q}_i = \sum_{i=1}^{n} \sum_{j=1}^{n} m_{ij} \dot{q}_j \dot{q}_i = 2T$$

The differential of this expression creates the term to be eliminated since

$$2dT = \sum_{i=1}^{n} d\left(\frac{\partial T}{\partial q_i}\right) \dot{q}_i + \sum_{i=1}^{n} \left(\frac{\partial T}{\partial \dot{q}_i}\right) d\dot{q}_i$$

Combining the two equations for dT to eliminate $\sum_{i=1}^{n} (\partial T / \partial \dot{q}_i) \, d\dot{q}_i$ produces a new expression for dT, which is

$$dT = \sum_{i=1}^{n} \left[d\left(\frac{\partial T}{\partial \dot{q}_i}\right) \dot{q}_i - \frac{\partial T}{\partial q_i} dq_i \right]$$

Since $\dot{q}_i = dq_i / dt$ and dt is a scalar quantity that can be shifted, as shown in the rearrangements performed in the next equation

$$d\left(\frac{\partial T}{\partial \dot{q}_i}\right) \dot{q}_i = d\left(\frac{\partial T}{\partial \dot{q}_i}\right) \frac{dq_i}{dt}$$

$$= \frac{d}{dt}\left(\frac{\partial T}{\partial \dot{q}_i}\right) dq_i$$

dT can be expressed in the form

$$dT = \sum_{i=1}^{n} \left[\frac{d}{dt}\left(\frac{\partial T}{\partial \dot{q}_i}\right) - \frac{\partial T}{\partial q_i} \right] dq_i$$

The differential of V is

$$dV = \sum_{i=1}^{n} \frac{\partial V}{\partial q_i} dq_i$$

so that the conservation equation for $T + V$ becomes

$$d(T + V) = \sum_{i=1}^{n} \left[\frac{d}{dt}\left(\frac{\partial T}{\partial \dot{q}_i}\right) - \frac{\partial T}{\partial q_i} + \frac{\partial V}{\partial q_i} \right] dq_i = 0$$

The generalized coordinates are independent of each other and, therefore, may be chosen arbitrarily. If we think of dq_i as components of the differential length of an n-dimensional vector, that vector may be assigned any orientation by the arbitrary choice of its components, dq_i. If we also imagine the coefficients

$$\frac{d}{dt}\left(\frac{\partial T}{\partial \dot{q}_i}\right) - \frac{\partial T}{\partial q_i} + \frac{\partial V}{\partial q_i}\right)$$

to be components of an n-dimensional vector, the conservation equation given before shows that these vectors are either always orthogonal to each other or that one of them is always a zero vector. Because the differential length vector can have any orientation, we can rule out orthogonality. Since it is also a vector of nonzero length, we can only conclude that the other vector is of zero length, which requires that all of its components be of zero length. As a consequence, the conservation equation becomes a set of scalar equations:

$$\frac{d}{dt}\left(\frac{\partial T}{\partial \dot{q}_i}\right) - \frac{\partial T}{\partial q_i} + \frac{\partial V}{\partial q_i} = 0 \qquad i = 1, 2, 3, \ldots, n$$

Because V is not a function of $\dot{q}_i$, then $\partial V/\partial \dot{q}_i = 0$. Subtracting this zero from the previous equations yields

$$\frac{d}{dt}\left(\frac{\partial T}{\partial \dot{q}_i} - \frac{\partial V}{\partial \dot{q}_i}\right) - \left(\frac{\partial T}{\partial q_i} - \frac{\partial V}{\partial q_i}\right) = 0 \qquad i = 1, 2, 3, \ldots, n$$

Now by defining the Lagrangian as $L = T - V$, the "conservative" version of the Lagrange equations reduces to

$$\frac{d}{dt}\left(\frac{\partial L}{\partial \dot{q}_i}\right) - \frac{\partial L}{\partial q_i} = 0 \qquad i = 1, 2, 3, \ldots, n$$

When the system is subjected to forces that cannot be represented by the gradient of a potential, the energy equation becomes

$$d(T + V) = dW$$

where dW is the work of those nonpotential forces that give rise to the system motion. The forces required for static equilibrium are obviously not included in dW. The differential work can be expressed in terms of the generalized forces Q_i and the generalized coordinates by

$$dW = \sum_{i=1}^{n} Q_i \, dq_i$$

The inclusion of these forces modifies the Lagrange equations to read

$$\frac{d}{dt}\left(\frac{\partial L}{\partial \dot{q}_i}\right) - \frac{\partial L}{\partial q_i} = Q_i \qquad i = 1, 2, 3, \ldots, n$$

Note that the generalized forces must do work on the system and the equations they belong to are determined by the differential displacements that they act through. Obviously, "workless" constraint forces do not contribute to the Q_i's of dW.

5

MACHINE DYNAMICS

5.1 INTRODUCTION

In the previous chapters, we have been concerned with the components of machines: linkages, rams, sliders, etc. In the last chapter, we developed a technique using the Lagrange equations that permit us to examine complex assemblies of these components, i.e., machines. In this chapter, we will employ these equations to investigate some of the phenomena commonly encountered in machine dynamics.

Most of us are aware of instances where devices assembled from well-designed components do not work quite as anticipated. The interaction between these otherwise acceptable components is, for some reason, incompatible. The result is an assembly whose performance is unacceptable under certain operating conditions. In this chapter, we will examine the causes for such behavior and some of the remedies.

Our investigations will lead us into a branch of mechanics called "vibrations." Here we will encounter two general categories of vibrations: forced and self-excited.

Forced vibrations result from stimuli external to the vibrating system. The frequency of the system's response to these stimuli is dependent on the frequency of the stimuli, usually proportional to it. Mathematically, the response of the system is the solution of the nonhomogeneous differential equations of motion. Large-amplitude (resonant) oscillations occur when the forcing frequencies are near or at the natural frequencies of the system, i.e., the frequencies at which it would prefer to vibrate. A machine that tries to operate at or near a resonant condition for any period of time is obviously in trouble.

The vibrations caused by unbalanced wheels on an automobile is an example of a forced vibration. The stimulus in this case is the rotation of the mass center about the center of rotation. The frequency and severity of the vibration change with the road speed. Balancing the wheels reduces the magnitude of the forcing function, but not its frequency.

Self-excited vibrations result from an alternating stimulus that is created and sustained by the motion itself. In this case, the system will vibrate without any external stimulus. Because the stimulus comes from within the system itself, rather than from outside, self-excited vibrations are difficult to anticipate or diagnose.

A common example of a self-sustained vibration is the fluttering of a venetian blind slat when a steady breeze blows over it. Energy to sustain the vibration is drawn from the steady wind, but the excitation of the slat is maintained by its own oscillation.

Mathematically, self-sustained oscillations result from the solution of the homogeneous differential equations of motion. Vibrations of this type are potentially dangerous especially if they become unstable.

In the case of either forced or self-sustained vibrations, the important system characteristic is its natural frequencies. Since these frequencies are built into the machine at the time of its construction, it is difficult to alter them once the machine is built. For this reason, it is important to determine the natural frequencies of new or modified designs before commencing fabrication.

5.2 THE RESPONSE OF A TWO-MASS SYSTEM TO A CONCENTRIC VERTICAL FORCE

We will begin our study of machine vibrations with an examination of the simplest nontrivial dynamic model possible. From this investigation, we will discover both a problem and a solution to a problem.

Figure 5.1 shows a two-mass system. Mass m_1 represents the machine's frame, which is considered to be rigid and supported on a linear spring k_1. The frame is excited by a sinusoidal force. This force is assumed to act through g_1.

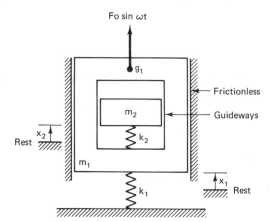

Figure 5.1 Two-Mass System with Sinusoidal Forcing

Mass m_2 represents a smaller mass contained within the machine. It is fastened elastically to the frame by linear spring k_2. The rest positions of m_1 and m_2 are the origins of the displacements x_1 and x_2, respectively. This system represents a simplified one-dimensional model of a machine mounted on elastic mounts with an elastically supported part.

The kinetic energy of the masses is

$$T = \tfrac{1}{2}m_1\dot{x}_1^2 + \tfrac{1}{2}m_2\dot{x}_2^2$$

The potential energy of the masses is

$$V = \tfrac{1}{2}k_1x_1^2 + \tfrac{1}{2}k_2(x_2 - x_1)^2$$

where the effect of gravity is omitted because the displacements of the masses are measured from their rest positions.

Since there are two generalized coordinates, x_1 and x_2, the Lagrange equations are

$$\frac{d}{dt}\left(\frac{\partial L}{\partial \dot{x}_1}\right) - \frac{\partial L}{\partial x_1} = Q_{x1}$$

$$\frac{d}{dt}\left(\frac{\partial L}{\partial \dot{x}_2}\right) - \frac{\partial L}{\partial x_2} = Q_{x2}$$

where the Lagrangian is

$$L = \tfrac{1}{2}m_1(\dot{x}_1)^2 + \tfrac{1}{2}m_2(\dot{x}_2)^2 - \tfrac{1}{2}k_1 x_1^2 - \tfrac{1}{2}k_2(x_2 - x_1)^2$$

and the generalized forces are

$$Q_{x1} = F_o \sin \omega t \qquad Q_{x2} = 0$$

where ω is the frequency of the exciting force.

The equations of motion that result are

$$m_1\ddot{x}_1 + k_1 x_1 - k_2(x_2 - x_1) = F_o \sin \omega t$$
$$m_2\ddot{x}_2 + k_2(x_2 - x_1) = 0 \tag{5.1}$$

These differential equations are coupled since they share the generalized coordinates x_1 and x_2. Coupled equations of motion are expected since the motion of either mass obviously influences the motion of the other through the coupling spring k_2.

The model does not include any damping elements. Some energy dissipation occurs during the flexure of "elastic" components due to hysteresis effects. This, combined with the frictional effects between moving parts, causes the starting transients to decay in times that are generally quite short compared to the total running time of most machines. As a consequence, steady-state operation is, by comparison, the dominant mode of operation. Unless special provisions are made to introduce dissipative elements, damping has little effect on the steady-state motion sustained by the forcing function.

The steady-state solutions of the coupled equations describing this problem are

$$x_1 = X_1 \sin \omega t \qquad x_2 = X_2 \sin \omega t$$

Inserting these into the equations of motion yields the algebraic equations

$$\left[1 + \frac{k_2}{k_1} - \left(\frac{\omega}{\omega_1}\right)^2\right]X_1 - \left(\frac{k_2}{k_1}\right)X_2 = X_o$$

$$-X_1 + \left[1 - \left(\frac{\omega}{\omega_2}\right)^2\right]X_2 = 0 \tag{5.2}$$

where

$$X_o = F_o/k_1 \qquad \omega_1 = \sqrt{k_1/m_1} \qquad \omega_2 = \sqrt{k_2/m_2}$$

The quantities ω_1 and ω_2 are the natural frequencies for each of the respective spring–mass combinations when each is mounted on a stationary base.

The solutions for the steady-state amplitudes are

$$\frac{X_1}{X_o} = \frac{1 - (\omega/\omega_2)^2}{\left[1 + \dfrac{k_2}{k_1} - \left(\dfrac{\omega}{\omega_1}\right)^2\right]\left[1 - \left(\dfrac{\omega}{\omega_2}\right)^2\right] - \dfrac{k_2}{k_1}}$$

$$\frac{X_2}{X_o} = \frac{1}{\left[1 + \dfrac{k_2}{k_1} - \left(\dfrac{\omega}{\omega_1}\right)^2\right]\left[1 - \left(\dfrac{\omega}{\omega_2}\right)^2\right] - \dfrac{k_2}{k_1}} \qquad (5.3)$$

The first of these equations shows that the displacement of the frame is zero when the excitation frequency ω equals the natural frequency of the mass m_2–spring k_2 system. At this frequency, the displacement of mass m_2 is

$$X_2 = -\frac{k_1}{k_2}X_o = -\frac{F_o}{k_2} \qquad \text{or} \qquad \frac{k_2 X_2}{F_o} = -1$$

The minus sign indicates that displacement of m_2 is 180° out of phase with the applied force.

Since x_1 is zero when $\omega = \omega_2$, x_2 is the total displacement of k_2. The force transmitted by spring k_2 is then $k_2 x_2 = k_2 X_2 \sin \omega_2 t$. The force applied to the system is $F_o \sin \omega_2 t$. The ratio of the transmitted force to the applied force is

$$\frac{k_2 X_2 \sin \omega_2 t}{F_o \sin \omega_2 t} = \frac{k_2 X_2}{F_o} = -1$$

where the minus sign indicates that the forces are oppositely directed. The equation shows that all of the applied force ($F_o \sin \omega_2 t$) is "absorbed" by moving mass m_2, leaving no force to excite frame m_1. At this frequency, mass m_2 acts as a vibration absorber, tuning out the natural frequency of frame mass m_1 and its supporting spring k_1.

This result shows that a troublesome natural frequency can be removed from an existing machine by adding a properly proportioned spring and mass to the system. Unfortunately, adding mass m_2 increases the number of degrees of freedom (generalized coordinates) to 2, which means that the number of natural frequencies of the new system is also increased to 2. Recall that ω_1 and ω_2 are the natural frequencies of the two individual spring–mass systems. They are not the resonant frequencies of the machine comprised of the two systems. The resonant frequencies of the two-mass system of Figure 5.1 are found by setting the denominator of the steady-state-amplitude equations equal to zero. The resulting algebraic equation is called the "characteristic equation" and its solution yields the resonant frequencies of the system.

A somewhat more direct method for obtaining this equation will be given shortly, along with an explanation of the rationale behind its formulation. By assuming for now that this is the correct way to proceed, the characteristic equation is

$$\left[1 + \frac{k_2}{k_1} - \left(\frac{\omega}{\omega_1}\right)^2\right]\left[1 - \left(\frac{\omega}{\omega_2}\right)^2\right] - \frac{k_2}{k_1} = 0 \qquad (5.4)$$

If we let $r = m_2/m_1$, then $k_2/k_1 = r(\omega_2/\omega_1)^2$, and the characteristic equation simplifies to

$$\left(\frac{\omega_2}{\omega_1}\right)^2\left(\frac{\omega}{\omega_2}\right)^4 - \left[1 + (1 + r)\left(\frac{\omega_2}{\omega_1}\right)^2\right]\left(\frac{\omega}{\omega_2}\right)^2 + 1 = 0$$

As an example of the effect an added mass can have, let the natural frequency of the added system be equal to the natural frequency of the original system. Then $\omega_1 = \omega_2$, and the characteristic equation simplifies to

$$\left(\frac{\omega}{\omega_1}\right)^4 - (2 + r)\left(\frac{\omega}{\omega_1}\right)^2 + 1 = 0$$

For practical reasons, one would not want to add a large mass to the original system, so that $r \ll 1.0$. In that case, the roots of the last equation are approximately

$$\frac{\omega}{\omega_1} \approx \left(1 + \frac{r}{2} \pm \sqrt{r}\right)^{1/2}$$

For example, if $r = 0.2$, then the system's resonant frequencies are $\omega = 0.8\omega_2$ and $\omega = 1.25\omega_2$, as shown in Figure 5.2(a). The location of the resonant frequencies versus the mass ratio is illustrated in Figure 5.2(b).

Obviously, the separation between the new resonant frequencies is not going to be large if the mass ratio is restricted to small values. Nevertheless, the added system does provide a relatively simple means of correcting resonance problems when the steady-state running frequency is fixed. Since the suppression of the original resonance has been attained at the expense of creating two new resonant frequencies,

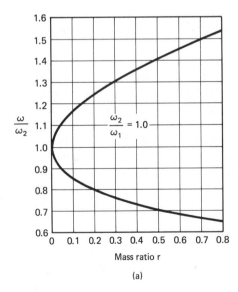

(a)

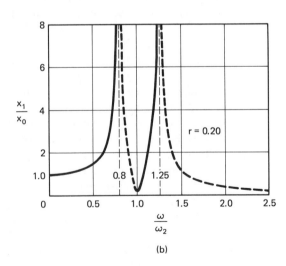

(b)

Figure 5.2   Resonant Frequencies of Two-Mass System

one on either side of the original resonance, care must be exercised during startup to avoid the destructive effects of the new resonant frequencies. This can be accomplished by passing rapidly through the lower frequency on the way to the running speed, ω, while at the same time avoiding a substantial overspeed to the higher frequency.

The result of this investigation can be considered from a different point of view. Suppose the vertical motion of a machine, thought to be relatively rigid, exhibits two natural frequencies instead of the expected one. If these frequencies bracket the expected frequency ω_1, they indicate that there is a flexibly supported mass attached to the machine that is excited by the forcing frequency. In real machines, there are usually a number of internal and external components that can vibrate, each giving rise to a resonant frequency. As a consequence, complex machines usually exhibit a number of amplitude "spikes" of varying magnitude spread out over the range of frequency excitation. Each of these spikes, or resonances, is the result of a particular mass making its contribution through one of its degrees of freedom. The example considered here is the simplest model of this phenomenon.

As noted earlier, there is a much more direct and more frequently used method for obtaining the characteristic equation that played such an important role in the analysis of this system's vibrational behavior. Since the characteristic equation "characterizes" the system and is independent of how the system is excited, it can be obtained directly from the homogeneous differential equations.

The solutions to the homogeneous differential equations

$$m_1 \ddot{x}_1 + k_1 x_1 - k_2(x_2 - x_1) = 0$$
$$m_2 \ddot{x}_2 + k_2(x_2 - x_1) = 0$$

are

$$x_1 = A_1(\sin \omega t + \phi_1) \qquad x_2 = A_2 (\sin \omega t + \phi_2)$$

These complementary solutions look very much like the particular solutions of the nonhomogeneous equations obtained earlier, but that is only because of the special form chosen for the forcing function, $f_a(t) = F_o \sin \omega t$. Regardless of the form of $f_a(t)$, the solutions to the homogeneous differential equations are the sinusoidal functions just given.

The simultaneous algebraic equations that result from applying the solutions to the differential equations are

$$\left[1 + \frac{k_2}{k_1} - \left(\frac{\omega}{\omega_1}\right)^2\right]A_1 - \left(\frac{k_2}{k_1}\right)A_2 = 0$$

$$-A_1 + \left[1 - \left(\frac{\omega}{\omega_2}\right)^2\right]A_2 = 0$$

Unlike the algebraic equations that result from the solution of the nonhomogeneous differential equations, these equations cannot be solved for the individual coefficients A_1 and A_2. At this point, we can only obtain equations for their ratio,

A_1/A_2, in terms of the as yet unknown frequencies ω. Of course, the same ratio A_1/A_2 should be obtained no matter which of the two equations is used to form that ratio. Since ω is the only "adjustable" parameter in these equations, we must conclude that a common ratio will be obtained only for specific values of ω, quantities that we will call the natural frequencies, ω_n, of the system.

Taking the previous equations and forming the ratio A_1/A_2 yields

$$\frac{A_1}{A_2} = \frac{k_2/k_1}{1 + k_2/k_1 - (\omega/\omega_1)^2}$$

and

$$\frac{A_1}{A_2} = \frac{1}{1 - (\omega/\omega_2)^2}$$

Equating the ratios and rearranging produces the equation:

$$\left[1 + \frac{k_2}{k_1} - \left(\frac{\omega}{\omega_1}\right)^2\right]\left[1 - \left(\frac{\omega}{\omega_2}\right)^2\right] - \frac{k_2}{k_1} = 0 \qquad (5.4)$$

which you will recognize as the equation for the resonant frequencies. Thus, we have discovered that a system's natural frequencies are also its resonant forcing frequencies. To put it another way, systems execute large-amplitude vibrations when forced to vibrate at their natural frequencies.

Once the natural (resonant) frequencies are found from this equation, ratios of A_1/A_2 associated with each of the natural frequencies can be calculated. The actual numerical values of these coefficients can only be assigned when the initial conditions are known.

For those who find it convenient to think geometrically, there is an interesting vector interpretation of the previous simultaneous algebraic equations. Combining these equations and writing them in matrix notation yields

$$\begin{bmatrix} 1 + \dfrac{k_2}{k_1} - \left(\dfrac{\omega}{\omega_1}\right)^2 & -\dfrac{k_2}{k_1} \\ -1 & 1 - \left(\dfrac{\omega}{\omega_2}\right)^2 \end{bmatrix}\begin{Bmatrix} A_1 \\ A_2 \end{Bmatrix} = \begin{Bmatrix} 0 \\ 0 \end{Bmatrix}$$

Defining the following two-dimensional vectors

$$\mathbf{R}_1 = \left[1 + \frac{k_2}{k_1} - \left(\frac{\omega}{\omega_1}\right)^2\right]\mathbf{i} + \left(-\frac{k_2}{k_1}\right)\mathbf{j}$$

$$\mathbf{R}_2 = (-1)\mathbf{i} + \left[1 - \left(\frac{\omega}{\omega_2}\right)^2\right]\mathbf{j}$$

$$\mathbf{A} = A_1\mathbf{i} + A_2\mathbf{j}$$

we find that the matrix equation asserts that the column vector $\mathbf{A}$ is orthogonal to each of the two row vectors, $\mathbf{R}_1$ and $\mathbf{R}_2$. It is clearly impossible for $\mathbf{A}$ to be simultaneously at right angles to two *different* vectors in the same two-dimensional space, so that we can only conclude that $\mathbf{R}_1$ and $\mathbf{R}_2$ are parallel to each other and differ only in their lengths. If that is the case, then their components are in the ratio

$$\frac{1 + k_2/k_1 - (\omega/\omega_1)^2}{-1} = \frac{-k_2/k_1}{1 - (\omega/\omega_2)^2}$$

Cross multiplication yields, once again, the characteristic equation obtained earlier, i.e.,

$$\left[1 + \frac{k_2}{k_1} - \left(\frac{\omega}{\omega_1}\right)^2\right]\left[1 - \left(\frac{\omega}{\omega_2}\right)^2\right] - \frac{k_2}{k_1} = 0 \tag{5.4}$$

We now see that the values of ω_n that satisfy the characteristic equation are those frequencies that cause the two row vectors of the matrix to be linearly related to each other by a scalar multiplier. (These frequencies are also called the eigenvalues.) For each ω_n, there is a vector $\mathbf{A}$ whose components are in a certain ratio that determine its orientation. These vectors, which are usually normalized in some convenient way, are called the eigenvectors. The lengths of the eigenvectors are determined by the initial conditions imposed on the problem. Each of these eigenvectors is orthogonal to the parallel row vectors of the matrix obtained by inserting the corresponding eigenvalue. We will have much more to say about this topic in later sections concerned with the vibrations of multidegree-of-freedom systems.

5.3 RESPONSE OF A RIGID MACHINE TO A VERTICAL EXCITATION

Figure 5.3(a) shows a rigid machine supported by two springs, k_1 and k_2, that maintain the machine level when at rest. The machine is constrained to move vertically and allowed to rock through angle θ, as illustrated in Figure 5.3(b). Coordinates X and Y are fixed in space, and coordinate y measures the vertical displacement of the

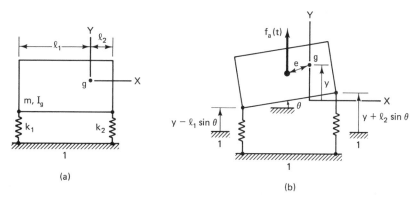

Figure 5.3   Rigid Machine on Elastic Supports

center of mass of the machine relative to its rest position. The applied vertical force $f_a(t)$ does not pass through the center of mass of the machine.

The kinetic energy of the machine is given by

$$T = \tfrac{1}{2}m(\dot{y})^2 + \tfrac{1}{2}I_g(\dot{\theta})^2$$

and the potential energy by

$$V = \tfrac{1}{2}k_1(y - l_1 \sin \theta)^2 + \tfrac{1}{2}k_2(y + l_2 \sin \theta)^2$$

The Lagrangian is, therefore,

$$L = T - V = \tfrac{1}{2}m(\dot{y})^2 + \tfrac{1}{2}I_g(\dot{\theta})^2 - \tfrac{1}{2}k_1(y - l_1 \sin \theta)^2 - \tfrac{1}{2}k_2(y + l_2 \sin \theta)^2$$

Coordinates y and θ are kinematically independent and are, therefore, chosen as generalized coordinates, so that the Lagrange equations for the system are

$$\frac{d}{dt}\left(\frac{\partial L}{\partial \dot{y}}\right) - \frac{\partial L}{\partial y} = Q_y$$

and

$$\frac{d}{dt}\left(\frac{\partial L}{\partial \dot{\theta}}\right) - \frac{\partial L}{\partial \theta} = Q_\theta$$

Since the force $f_a(t)$ does not have the displacement y, the equivalent force-couple system shown in Figure 5.4 must be constructed. From this, we find that the generalized force and couple are

$$Q_y = f_a(t)$$

$$Q_\theta = -f_a(t)e \cos \theta$$

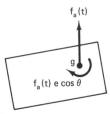

f$_a$(t)

g

f$_a$(t) e cos θ

Figure 5.4 Equivalent Force–Couple System

With these inputs, the Lagrange equations yield the differential equations of motion:

$$m\ddot{y} + k_1(y - l_1 \sin \theta) + k_2(y + l_2 \sin \theta) = f_a(t)$$

$$I_g\ddot{\theta} + k_1(y - l_1 \sin \theta)(-l_1 \cos \theta) + k_2(y + l_2 \sin \theta)(l_2 \cos \theta) = -f_a(t)e \cos \theta$$

$$(5.5)$$

If the machine has a reasonably well-designed support system, the angular displacement θ will not be very large. In that case, $\sin \theta \cong \theta$ and $\cos \theta \cong 1.0$. The resulting linearized equations of motion for small θ are

$$m\ddot{y} + (k_1 + k_2)y + (k_2 l_2 - k_1 l_1)\theta = f_a(t)$$
$$I_g\ddot{\theta} + (k_2 l_2 - k_1 l_1)y + (k_1 l_1^2 + k_2 l_2^2)\theta = -f_a(t)e \tag{5.6}$$

Notice that these equations are coupled and that the coupling disappears when $k_2 l_2 = k_1 l_1$. There are two degrees of freedom and each will yield a natural frequency of the system. Based on the previous example, we expect these frequencies to be dependent on the natural frequencies associated with each of the two modes of vibration, i.e., the translational vertical motion and the angular rocking motion.

To simplify the solution of these equations, we will define the following coefficients

$$a = (1/m)(k_1 + k_2)$$
$$b = (1/m)(k_2 l_2 - k_1 l_1)$$
$$c = (1/I_g)(k_1 l_1^2 + k_2 l_2^2)$$
$$d = b(m/I_g)$$

The equations of motion can now be written in the condensed form:

$$\ddot{y} + ay + b\theta = f_a(t)/m$$
$$\ddot{\theta} + c\theta + dy = -f_a(t)e/I_g \tag{5.7}$$

Ordinarily, the exact nature of the forcing function $f_a(t)$ is unknown. In the previous section, it was assumed to be sinusoidal to illustrate the form of both the forced and free response. Fortunately, the resonant conditions, i.e., the natural frequencies, of the system can be determined independent of excitation. To obtain the characteristic equation and the natural frequencies, we only need to know the form of the solution to the homogeneous differential equation. In the last section, we found that if the forcing function contains any of the system's natural frequencies, the system will greatly amplify the excitation amplitudes at these frequencies. This is certainly something we would always like to avoid.

For the differential equations at hand, the form of the homogeneous solutions will be taken to be

$$y = Ye^{\omega t} \qquad \theta = \Theta e^{\omega t}$$

These assumptions appear to be quite different from the sinusoids chosen in the previous section. As we will see, these exponentials reduce to pure sinusoids because the ω_n's obtained as roots of the characteristic equation are purely imaginary. Exponential solutions are introduced here to prepare the way for their eventual use in more involved problems.

When the assumed exponential solutions are inserted into the coupled differential equations, they yield the following coupled algebraic equations for the amplitudes Y and Θ:

$$Y\omega^2 + aY + b\Theta = 0$$
$$\Theta\omega^2 + c\Theta + dY = 0$$

or

$$(\omega^2 + a)Y + b\Theta = 0$$

$$dY + (\omega^2 + c)\Theta = 0$$

The characteristic equation is, therefore,

$$\begin{vmatrix} \omega^2 + a & b \\ d & \omega^2 + c \end{vmatrix} = 0$$

or

$$\omega^4 + (a + c)\omega^2 + ac - bd = 0$$

Since this equation is in the quadratic form, the roots are

$$\omega_{1,2}^2 = -\frac{a + c}{2} \pm \sqrt{\left(\frac{a - c}{2}\right)^2 + bd}$$

It can be shown that ω_1^2 and ω_2^2 are real negative numbers, so that

$$\omega_1 = \pm i \sqrt{\frac{a + c}{2} - \sqrt{\left(\frac{a - c}{2}\right)^2 + bd}} = \pm i\omega_{n1}$$

$$\omega_2 = \pm i \sqrt{\frac{a + c}{2} - \sqrt{\left(\frac{a - c}{2}\right)^2 + bd}} = \pm i\omega_{n2}$$

(5.8)

The $\pm i$ simply means that y and θ are sinusoids, with frequencies given by the coefficients of $i = \sqrt{-1}$.

Because these expressions are complicated, we will investigate several special cases. If the coupling were completely absent, then $b = 0 = d$, so that the uncoupled natural frequencies are $\omega_{n1} = \sqrt{c}$ and $\omega_{n2} = \sqrt{a}$. (These results can also be found by inspection of the linearized differential equations with $b = d = 0$.) These are the natural frequencies associated with purely translational and rotational vibrations.

If the coupling coefficient is now assumed to be small so that

$$bd \ll \left(\frac{a - c}{2}\right)^2$$

then the "lightly" coupled natural frequencies of the system are

$$\omega_{n1} \cong \sqrt{c - \left(\frac{bd}{a - c}\right)} \qquad \omega_{n2} \cong \sqrt{a + \left(\frac{bd}{a - c}\right)}$$

These approximations show that the coupled natural frequencies lie outside the uncoupled natural frequencies, i.e., above the largest and below the smallest. For larger coupling coefficients, this separating trend continues.

At this point in the analysis, we have already determined most of what is interesting to the designer, i.e., the natural frequencies of the system and how they depend on the system's parameters. With this information, the designer can decide whether any changes need be made in the design and what constraints, if any, need be placed on the operating-speed range to avoid resonance.

For the sake of completeness, we will conclude this section with an analysis of the steady-state response to a sinusoidal forcing function $f_a(t) = F_o \sin \omega t$.

By assuming that steady-state solutions are $y = B \sin \omega t$ and $\theta = D \sin \omega t$, the linearized differential equations reduce to

$$(a - \omega^2)B + (b)D = F_o/m$$

$$(d)B + (c - \omega^2)D = -F_o e/I_g$$

The solutions for B and D are

$$B = F_o\left[\frac{(c - \omega^2)/m + be/I_g}{\omega^4 - (a + c)\omega^2 + ac - bd}\right]$$

$$D = -F_o\left[\frac{(a - \omega^2)e/I_g + b/I_g}{\omega^4 - (a + c)\omega^2 + ac - bd}\right]$$

(5.9)

The denominator of each of these coefficients set equal to zero is the characteristic equation for the natural frequency, ω_n, as shown before.

Although the damping mechanisms that cause the transient solution to decay have been omitted from the analysis, we know that in real systems decay will occur. Since the coefficients of the decaying transient response are determined by the initial conditions, their effect on the complete solution will also disappear in time. For this reason, linear systems eventually "forget" how they started.

The effects of damping are included in the analysis when we wish to examine the starting transients. For very large machines that start rather slowly, the earlier responses of the machine should be carefully considered.

5.4 RESPONSE OF A RIGID MACHINE TO AN UNBALANCED ROTOR

Equations of Motion

Figure 5.5(a) shows a machine whose base is constrained by horizontal and vertical springs. An unbalanced rotor m_2 with its center of mass at g_2 is rigidly mounted within the frame of the machine. The center of mass of the frame, m_1, is assumed to be at the geometric center of the rotor, i.e., on the rotor's center line.

To evaluate the kinetic energy of masses m_1 and m_2, a fixed reference frame XY and a moving reference frame xy will be employed. The xy reference frame is fixed to the frame of the machine, with its origin at g_1. Coordinates xy locate the center of mass of rotor m_2 relative to the machine's frame. The angular rotation of this reference frame is the angular rotation of the machine's frame. The coordinates XY locate g_1, the center of mass of the machine's frame relative to ground.

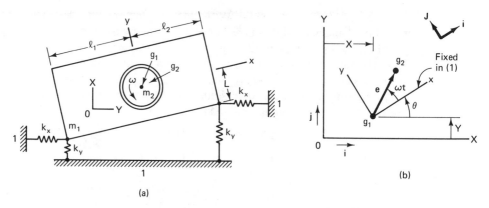

Figure 5.5   Rigid Machine with Unbalanced Rotor

It will be assumed that the center of mass g_2 rotates about g_1 at a fixed distance e. The angular velocity of the rotor relative to the frame is ω, so that the components of the vector $\mathbf{e}$ in the xy system are

$$x = e \cos \omega t \qquad y = e \sin \omega t$$

It will also be assumed that ω is constant.

The total kinetic energy of the frame is

$$T_1 = \tfrac{1}{2} m_1 V_{g1}^2 + \frac{1}{2} I_{g1} \dot{\theta}^2 = \tfrac{1}{2} m_1 (\dot{X}^2 + \dot{Y}^2) + \frac{1}{2} I_{g1} \dot{\theta}^2$$

To formulate the kinetic energy of the rotor, it will be necessary to formulate the velocity of the center of mass of the rotor, g_2. Applying the basic velocity equation yields:

$$\mathbf{V}_{g2} = \mathbf{V}_{g1} + \dot{\boldsymbol{\theta}} \times \mathbf{e} + \mathbf{v}_{xyz}$$

where

$\mathbf{V}_{g1} = \dot{X}\mathbf{I} + \dot{Y}\mathbf{J} = $ velocity of the xy origin
$\dot{\boldsymbol{\theta}} = \dot{\theta}\mathbf{k} = $ angular velocity of the xy frame
$\mathbf{e} = (e \cos \omega t)\mathbf{i} + (e \sin \omega t)\mathbf{j} = $ relative position vector of g_2
$\mathbf{v}_{xyz} = \dot{\mathbf{e}} = -\omega e[(\sin \omega t)\mathbf{i} - (\cos \omega t)\mathbf{j}]$

Inserting these into the velocity equation yields

$$\mathbf{V}_{g2} = \dot{X}\mathbf{I} + \dot{Y}\mathbf{J} + e(\omega + \dot{\theta})[(-\sin \omega t)\mathbf{i} + (\cos \omega t)\mathbf{j}]$$

In this equation, velocity $\mathbf{V}_{g2}$ is expressed in terms of a mixed set of base vectors, i.e., $\mathbf{i}$, $\mathbf{j}$, and $\mathbf{I}$, $\mathbf{J}$. Before constructing the kinetic energy of g_2, $\mathbf{V}_{g2}$ will be formulated in terms of a common set of base vectors. Since

$$\mathbf{i} = (\cos \theta)\mathbf{I} + (\sin \theta)\mathbf{J}$$

$$\mathbf{j} = -(\sin \theta)\mathbf{I} + (\cos \theta)\mathbf{J}$$

the velocity vector can be expressed in the fixed coordinate system as

$$\mathbf{V}_{g2} = [\dot{X} - e(\omega + \dot{\theta}) \sin (\theta + \omega t)]\mathbf{I} + [\dot{Y} + e(\omega + \dot{\theta}) \cos (\theta + \omega t)]\mathbf{J}$$

so that the square of the magnitude of the velocity vector is

$$V_{g2}^2 = [\dot{X} - e(\omega + \dot{\theta}) \sin (\theta + \omega t)]^2 + [\dot{Y} + e(\omega + \dot{\theta}) \cos (\theta + \omega t)]^2$$

Notice that in this analysis, the relative angular velocity of the rotor with respect to the frame, ω, has combined with the absolute angular velocity of the frame with respect to the ground, $\dot{\theta}$, to form the absolute velocity of the rotor with respect to the ground, $\omega + \dot{\theta}$.

The same thing has happened with regard to the rotor's angular displacement, i.e., $\theta + \omega t$ is the total angular displacement of the rotor with respect to the ground. Ordinarily, the angular displacement of the frame, θ, is very small compared to ωt, especially for large t. The same cannot be said for the relative sizes of ω and $\dot{\theta}$, although in the case of relatively massive machines, one would not expect the machine to rock at high frequencies even when the rotor is turning rapidly.

The total kinetic energy of mass m_2 can now be expressed as

$$T_2 = \tfrac{1}{2}m_2\{[\dot{X} - e(\omega + \dot{\theta}) \sin (\theta + \omega t)]^2 + [\dot{Y} + e(\omega + \dot{\theta}) \cos (\theta + \omega t)]^2\}$$
$$+ \tfrac{1}{2}I_{g2}(\omega + \dot{\theta})^2$$

As usual, the potential energy will be measured from the static equilibrium position, which will be assumed to be $X = Y = \theta = 0$. Only the contributions of the springs need be considered, so that

$$V = \tfrac{1}{2}k_y(Y - l_1 \sin \theta)^2 + \tfrac{1}{2}k_y(Y + l_2 \sin \theta)^2 + 2(\tfrac{1}{2})k_x(X + L \sin \theta)^2$$

When the Lagrangian $L = T_1 + T_2 - V$ is formed, it will contain three generalized coordinates. They are X, Y, and θ, so that there will be three Lagrange equations. Since the forces and couples that cause the machine to translate and rotate come from within the machine, i.e., the unbalance of the rotor, the generalized forces and couples associated with these equations will all be zero.

The first Lagrange equation is

$$\frac{d}{dt}\left(\frac{\partial L}{\partial \dot{X}}\right) - \frac{\partial L}{\partial X} = Q_x$$

which yields

$$(m_1 + m_2)\ddot{X} + 2k_x(X + L \sin \theta) = m_2 e[\ddot{\theta} \sin (\theta + \omega t)$$
$$+ (\omega + \dot{\theta})^2 \cos (\theta + \omega t)] \quad (5.10a)$$

The second Lagrange equation is

$$\frac{d}{dt}\left(\frac{\partial L}{\partial \dot{Y}}\right) - \frac{\partial L}{\partial Y} = Q_Y$$

which yields

$$(m_1 + m_2)\ddot{Y} + 2k_y Y + k_y(l_2 - l_1)\sin\theta$$
$$= m_2 e[-\ddot{\theta}\cos(\theta + \omega t) + (\omega + \dot{\theta})^2 \sin(\theta + \omega t)]$$

$$(5.10b)$$

These equations have been arranged so that the forcing or exciting terms are on the right-hand sides of each equation. Each of these forcing functions is proportional to $m_2 e$, showing that the source of the vibration diminishes as $m_2 e$ approaches zero. As expected, the machine's vibrations vanish when its excitation vanishes. Obviously, one would want to reduce e as much as practical to improve the balance.

The third Lagrange equation is

$$\frac{d}{dt}\left(\frac{\partial L}{\partial \dot{\Theta}}\right) - \frac{\partial L}{\partial \Theta} = Q_\Theta$$

which results in

$$J_1 \ddot{\theta} + k_y(Y + l_2 \sin\theta)(l_2 \cos\theta) - k_y(Y - l_1 \sin\theta)(l_1 \cos\theta)$$
$$+ 2k_x(X + L\sin\theta)(L\cos\theta) = m_2 e[\ddot{X}\sin(\theta + \omega t) - \ddot{Y}\cos(\theta + \omega t)]$$

$$(5.10c)$$

where $J_1 = I_{g1} + (I_{g2} + e^2 m_2)$, the polar moment of inertia of the whole machine about g_1.

It is always good practice to pause at this stage of the analysis to examine the physical significance of each of the terms that appear in the equations. If there are errors or omissions, they can often be detected at this point.

The first terms on the left-hand side of each of these equations are the inertial effects of translation and rotation. The remaining terms, all containing the spring constants, are the forces and moments that the machine supports exert on the frame.

The exciting forces on the right-hand sides of the first two equations have a simple geometric interpretation, as illustrated in Figure 5.6.

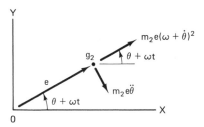

Figure 5.6 Unbalance Inertia Forces

From this figure, the X and Y components of the d'Alembert inertial forces due to the unbalanced e are found to be the same as the right-hand sides of these equations, i.e.,

$$m_2 e[(\omega + \dot{\theta})^2 \cos(\theta + \omega t) + \ddot{\theta}\sin(\theta + \omega t)]$$

and

$$m_2 e[(\omega + \dot\theta)^2 \sin(\theta + \omega t) - \ddot\theta \cos(\theta + \omega t)]$$

The exciting moment on the right-hand side of the last equation can be obtained from Figure 5.7. The inertial forces in this moment result from the motion of the frame rather than the rotor as in the first two force equations.

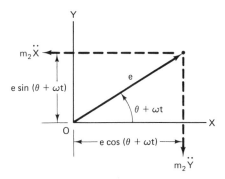

Figure 5.7 Unbalance Inertia Moment

The differential equations derived are nonlinear and cannot be solved by ordinary techniques. Assuming that θ is small permits the approximations $\sin\theta \cong \theta$, $\cos\theta \cong 1$, and $\omega t + \theta \cong \omega t$. If we make the further assumption that the rotor speed ω is much larger than the rocking velocity of the machine, the equations can be linearized. They then become

$$(m_1 + m_2)\ddot X + 2k_x X + 2k_x L\theta = m_2 e(\ddot\theta \sin\omega t + \omega^2 \cos\omega t)$$

$$(m_1 + m_2)\ddot Y + 2k_y Y + k_y(l_2 - l_1)\theta = m_2 e(-\ddot\theta \cos\omega t + \omega^2 \sin\omega t) \qquad (5.11)$$

$$J_1\ddot\theta + k_y(l_2 - l_1)Y + [k_y(l_2^2 + l_1^2) + 2k_x L^2]\theta + 2k_x LX$$
$$= m_2 e(\ddot X \sin\omega t - \ddot Y \cos\omega t)$$

If $k_x = 0$ and $k_y = k_1 = k_2$, the left-hand sides of the last two differential equations reduce to the left-hand sides from the linearized differential equations of the previous section. The first equation accounts for the additional horizontal degree of freedom not permitted in the problem of Section 5.3.

Characteristic Equation

The complete solution of the linearized equations is still difficult due to the presence of the dependent variables on the right-hand side of the equations, where they are multiplied by the trigonometric functions of time.

Fortunately, much useful information about the behavior of the system can be gained by investigating the homogeneous differential equations that remain when the excitation, unbalance e, is removed. As before, we will look to the characteristic equation and the natural frequencies it reveals. To simplify the analysis, let

$$a = \frac{2k_y}{m_1 + m_2} \qquad c = \frac{k_y(l_1^2 + l_2^2) + 2k_x L^2}{J_1}$$

$$b = \frac{k_y(l_2 - l_1)}{m_1 + m_2} \qquad q = \frac{2k_x}{m_1 + m_2}$$

The homogeneous differential equations can now be written

$$\ddot{X} + qX + Lq\theta \approx 0$$

$$\ddot{Y} + aY + b\theta \approx 0 \qquad (5.12)$$

$$\ddot{\theta} + \frac{b}{J_1}(m_1 + m_2)Y + c\theta + \frac{qL}{J_1}(m_1 + m_2)X = 0$$

In these equations, b and q are the coupling coefficients. The characteristic equation is found, as before, by assuming exponential solutions for X, Y, and θ. The solution of the characteristic equation in this case yields three natural frequencies, one for each degree of freedom.

The characteristic equation is

$$\begin{vmatrix} \omega^2 + q & 0 & Lq \\ 0 & \omega^2 + a & b \\ \dfrac{(m_1 + m_2)Lq}{J_1} & \dfrac{b(m_1 + m_2)}{J_1} & \omega^2 + c \end{vmatrix} = 0 \qquad (5.13)$$

Resonance is expected if the steady-state response occurs at any of the system's natural frequencies. The steady-state response depends on the excitation and is, as noted earlier, the solution of the nonhomogeneous differential equations. Since the forcing functions of all three differential equations are periodic, we would expect that the solutions X, Y, and θ also to be periodic. The question is whether the frequency of these solutions is at or near the natural frequencies of the system, and if so at what shaft running speeds, ω.

Perturbation Solution for Small Unbalance

As already indicated, the solution of these nonhomogeneous linearized differential equations is not straightforward. Some further approximations will obviously be necessary to obtain any indication of what the steady-state response looks like. To determine what further simplifications can be made, we will need to gain some further insight into the nature of the solution. The next sections will develop a rational basis for these assumptions and what they lead to.

We will begin the development of our approximate homogeneous solution with the observation that the solutions for X and Y are lengths and, therefore, they must be proportional to some characteristic length of the system. Since the lengths X and Y vanish when the unbalance length e vanishes, they must be proportional to the length e. Even in a poorly balanced machine, $m_2 e$ is of the order of several inch-ounces, so that e is usually very small, in the order of thousandths of an inch.

Based on this observation, we will define two dimensionless displacements as $\bar{X} = X/e$ and $\bar{Y} = Y/e$, where the magnitudes of the dimensionless dependent variables $\bar{X}$ and $\bar{Y}$ are of the order of unity. Applying these to the right-hand side of the moment equation (divided by J_1) yields

$$\frac{m_2 e}{J_1} (\ddot{X} \sin \omega t - \ddot{Y} \cos \omega t) = \frac{m_2 e^2}{J_1} (\ddot{\bar{X}} \sin \omega t - \ddot{\bar{Y}} \cos \omega t)$$

The magnitude of this forcing function is of the order $m_2 e^2/J_1$ since the terms in parentheses are of the order of unity. The ratio $m_2 e^2/J_1$ is a very small number, which suggests that as a first approximation, the right-hand side of the moment equation could be treated as approximately zero, which removes the moment excitation from its right-hand side, reducing it to a homogeneous differential equation.

Unfortunately, there is no characteristic angle in the problem, but we do know that θ vanishes when e vanishes. Since θ is dimensionless, it must be proportional to a length *ratio* such as e/L. With this observation, we can define $\theta = (e/L)\bar{\theta}$, so that e cancels out of the now homogeneous moment equation.

When this scaling is applied to the right-hand sides of the first two equations, they become:

$$m_2 e (\ddot{\theta} \sin \omega t + \omega^2 \cos \omega t) = m_2 e \left(\frac{e}{L} \ddot{\bar{\theta}} \sin \omega t + \omega^2 \cos \omega t \right) \cong m_2 e \omega^2 \cos \omega t$$

and

$$m_2 e (-\ddot{\theta} \cos \omega t + \omega^2 \sin \omega t) = m_2 e \left(-\frac{e}{L} \ddot{\bar{\theta}} \cos \omega t + \omega^2 \sin \omega t \right) \cong m_2 e \omega^2 \sin \omega t$$

The simplifications made here reflect the fact that $(e/L)\theta$ can be made arbitrarily small compared to ω by adjusting e. The linearized equations with their approximate forcing functions (for $e \rightarrow 0$) now become

$$\ddot{\bar{X}} + q\bar{X} + q\bar{\theta} = \left(\frac{m_2 \omega^2}{m_1 + m_2} \right) \cos \omega t$$

$$\ddot{\bar{Y}} + a\bar{Y} + (b/L)\bar{\theta} = \left(\frac{m_2 \omega^2}{m_1 + m_2} \right) \sin \omega t \qquad (5.14)$$

$$\ddot{\bar{\theta}} + c\bar{\theta} + \left[\frac{bL(m_1 + m_2)}{J_1} \right] \bar{Y} + \left[\frac{L^2 q (m_1 + m_2)}{J_1} \right] \bar{X} = 0$$

The first two approximate equations indicate that when e is small, the forcing function for the linear displacements X and Y is essentially the centrifugal force created by the unbalance of the rotor, as we might have assumed based on intuition. The third equation indicates that the rocking motion is largely caused by the linear displacement of the springs at the base of the machine.

The steady-state solutions of these equations contain sin ωt and cos ωt, so that resonance will occur when the rotor running speed ω is approximately equal to any one of the three natural frequencies obtained from the characteristic equation. The validity of this conclusion improves as the unbalance is decreased.

The result obtained is called a perturbation solution. The unbalance distance e is the perturbing parameter, which gives rise to the perturbations in X, Y, and θ. The perturbing relations used before were developed from plausible physical arguments. For small values of e, we assumed that there is a linear relationship between the cause, e, and the effects, X, Y, and θ. The responses obtained from the resulting simplified differential equations, although limited to small values of the unbalance, indicate roughly what we can expect from this complicated dynamic system as we improve the balance.

The perturbation technique can be applied in a formal way by assuming that the dependent variables can be expressed in the form of a series by

$$X = eX_1 + e^2X_2 + \cdots$$
$$Y = eY_1 + e^2Y_2 + \cdots$$
$$\theta = e\Theta_1 + e^2\Theta_2 + \cdots \tag{5.15}$$

where $X_1, X_2, \ldots, Y_1, Y_2, \ldots$, and $\Theta_1, \Theta_2, \ldots$ are functions of time to be found, and e is a small parameter. When these assumed solutions are inserted into the original differential equations, they generate a sequence of differential equations that can then be solved in order for X_1, X_2, etc. For example, the linearized differential equation that results from the first Lagrange equation becomes

$$(m_1 + m_2)[e\ddot{X}_1 + e^2\ddot{X}_2 + \cdots] + 2k_x[eX_1 + e^2X_2 + \cdots]$$
$$+ 2k_xL[e\Theta_1 + e^2\Theta_2 + \cdots]$$
$$= m_2e[(e\ddot{\Theta}_1 + e^2\ddot{\Theta}_2 + \cdots)\sin \omega t + \omega^2 \cos \omega t]$$

Since this equation must be satisfied for all values of e, we can form a sequence of differential equations by equating the coefficients of like powers of e on both sides of the previous equation. From the coefficients of e, we get

$$(m_1 + m_2)\ddot{X}_1 + 2k_xX_1 + 2k_xL\Theta_1 = m_2\omega^2 \cos \omega t$$

which is equivalent to the first of the perturbation differential equations obtained earlier.

Similarly, the other two differential equations yield

$$(m_1 + m_2)\ddot{Y}_1 + 2k_yY_1 + k_y(l_2 - l_1)\Theta_1 = m_2\omega^2 \sin \omega t$$
$$J_1\ddot{\Theta}_1 + k_y(l_2 - l_1)Y_1 + [k_y(l_2^2 + l_1^2) + 2k_xL^2]\Theta_1 + 2k_xLX_1 = 0$$

The simultaneous solution of these equations for X_1, Y_1, and Θ_1 can then be inserted into the right-hand side of the differential equations that result from the coefficients of e^2. These equations then become

$$(m_1 + m_2)\ddot{X}_2 + 2k_x X_2 + 2k_x L\Theta_2 = m_2 \ddot{\Theta}_1 \sin \omega t$$

$$(m_1 + m_2)\ddot{Y}_2 + 2k_y Y_2 + k_y(l_2 - l_1)\Theta_2 = -m_2 \ddot{\Theta}_1 \cos \omega t$$

$$J_1\ddot{\Theta}_2 + k_y(l_2 - l_1)Y_2 + [k_y(l_2^2 + l_1^2) + 2k_x L^2]\theta_2 + 2k_x LX_2$$
$$= m_2(\ddot{X}_1 \sin \omega t - \ddot{Y}_1 \cos \omega t)$$

These equations are then solved for X_2, Y_2, and Θ_2.

Although laborious, this process can be continued (at least in principle) to any power-of-e approximation desired. In practice, approximations higher than quadratic in e are rarely calculated.

Perturbation solutions are frequently used to obtain approximate solutions to difficult problems containing a small parameter. They also provide a means of checking the validity of numerical solutions obtained from computer codes.

5.5 TORSIONAL VIBRATIONS

Many machine rotors consist of a shaft supporting one or more relatively rigid disks, such as shown in Figure 5.8. In this system, the kinetic energy is stored primarily in the disks, and the potential energy is stored in the torsional elasticity of the shaft sections between the disks.

Figure 5.9 shows the lumped-parameter model of this system, where k_{o1} and k_{12} are torsional spring equivalents of the shaft sections. For each section of shaft, the stiffness would be calculated from the equation

$$k = \frac{GJ}{l}$$

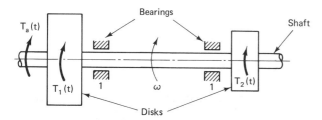

Figure 5.8   Rotor with Two Disks

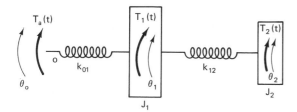

Figure 5.9   Lumped Model of Two-Disk Rotor

where G is the shear modulus, J is the polar moment of the shaft cross section, and l is the length of the shaft section between disks.

The total kinetic energy of the rotor system is

$$T = \tfrac{1}{2} J_1 (\dot{\theta}_1)^2 + \tfrac{1}{2} J_2 (\dot{\theta}_2)^2$$

and the potential energy is

$$V = \tfrac{1}{2} k_{o1} (\theta_o - \theta_1)^2 + \tfrac{1}{2} k_{12} (\theta_1 - \theta_2)^2$$

where the angles are measured from rest.

The Lagrangian is

$$L = T - V = \tfrac{1}{2} J_1 (\dot{\theta}_1)^2 + \tfrac{1}{2} J_2 (\dot{\theta}_2)^2 - \tfrac{1}{2} k_{o1} (\theta_o - \theta_1)^2 - \tfrac{1}{2} k_{12} (\theta_1 - \theta_2)^2$$

Since θ_o, θ_1, and θ_2 are kinematically independent, they can be chosen as the generalized coordinates for three Lagrange equations. The first equation is

$$\frac{d}{dt} \left(\frac{\partial L}{\partial \dot{\theta}_o} \right) - \frac{\partial L}{\partial \theta_o} = Q_{\theta_o}$$

which becomes

$$k_{o1}(\theta_o - \theta_1) = T_a(t)$$

Similarly, the second and third equations become

$$J_1 \ddot{\theta}_1 + k_{o1}(\theta_1 - \theta_o) + k_{12}(\theta_1 - \theta_2) = T_1(t)$$
$$J_2 \ddot{\theta}_2 + k_{12}(\theta_2 - \theta_1) = T_2(t)$$

Substituting the first equation into the second reduces the differential equations to two, i.e.,

$$\begin{aligned}
J_1 \ddot{\theta}_1 + k_{12}(\theta_1 - \theta_2) &= T_1(t) + T_a(t) \\
J_2 \ddot{\theta}_2 + k_{12}(\theta_2 - \theta_1) &= T_2(t)
\end{aligned} \tag{5.16}$$

Once again, we have coupled differential equations. Since they are linear, the formation of the characteristic equation and the determination of the natural frequencies is straightforward. If the steady-state response contains any of these natural frequencies, then resonance will occur.

When the two disks are identical, or nearly so, the equations can be easily uncoupled. If the equations are added:

$$J(\ddot{\theta}_1 + \ddot{\theta}_2) = T_1(t) + T_2(t) + T_a(t)$$

If they are subtracted:

$$J(\ddot{\theta}_1 - \ddot{\theta}_2) + 2k_{12}(\theta_1 - \theta_2) = T_1(t) - T_2(t) + T_a(t)$$

Now defining new dependent variables

$$\theta_1 + \theta_2 = \phi_1$$

$$\theta_1 - \theta_2 = \phi_2$$

the differential equations become uncoupled, taking the form:

$$J\ddot{\phi}_1 = T_1(t) + T_2(t) + T_a(t)$$

$$J\ddot{\phi}_2 + 2k_{12}\phi_2 = T_1(t) - T_2(t) + T_a(t)$$
(5.17)

The solutions of the homogeneous equations

$$J\ddot{\phi}_1 = 0$$

$$J\ddot{\phi}_2 + 2k_{12}\phi_2 = 0$$

are

$$\phi_1(t) = C_{11} + C_{12}t$$

$$\phi_2(t) = C_{21} \sin\left(\sqrt{\frac{2k_{12}}{J}}t\right) + C_{22} \cos\left(\sqrt{\frac{2k_{12}}{J}}t\right)$$

Although the solution for $\phi_1(t)$ may seem a little strange at first glance, we will find there is a logical explanation for why it does not contain the sinusoidal functions we expect and actually have obtained for $\phi_2(t)$.

The solutions for ϕ_1 and ϕ_2 and, therefore, for θ_1 and θ_2 show that there are two natural frequencies, zero and $(2k_{12}/J)^{1/2}$. The "zero" natural frequency can be explained by considering the situation where the initial conditions for θ_1 and θ_2 are equal, i.e., $\theta_1(0) = \theta_2(0) \neq 0$ so that $\theta_1(0) - \theta_2(0) = 0$. Physically, this means that the two disks have both been rotated through a common angle and no elastic energy has been stored in spring k_{12}. This is called a "rigid-body rotation" since part of the disk–spring system has rotated as if it were a rigid body. Assuming that the disks were given equal initial velocities when they were placed in this new position, the whole system would rotate, but no oscillations would occur.

Suppose, instead, that the two disks were initially rotated through the same angle, but then released with different initial angular velocities. The initial angular rotation only changes the reference position for measuring the angles, i.e., it sets a new zero reference for measuring θ_1 and θ_2. Let θ_1 and θ_2 be zero following the initial rigid-body rotation. The initial conditions are now

$$\theta_1(0) = \theta_2(0) = 0$$

$$\dot{\theta}_1(0) \neq \dot{\theta}_2(0) \neq 0$$

The solutions now become

$$\phi_1(t) = [\dot{\theta}_1(0) + \dot{\theta}_2(0)]t$$

$$\phi_2(t) = \left[\frac{\dot{\theta}_1(0) - \dot{\theta}_2(0)}{\sqrt{2k_{12}/J}}\right] \sin\left(\sqrt{\frac{2k_{12}}{J}}t\right)$$

or

$$\theta_1(t) = \frac{1}{2}[\dot{\theta}_1(0) + \dot{\theta}_2(0)]t + \left[\frac{\dot{\theta}_1(0) - \dot{\theta}_2(0)}{2\sqrt{2k_{12}/J}}\right] \sin\left(\sqrt{\frac{2k_{12}}{J}}t\right)$$

$$\theta_2(t) = \frac{1}{2}[\dot{\theta}_1(0) + \dot{\theta}_2(0)]t - \left[\frac{\dot{\theta}_1(0) - \dot{\theta}_2(0)}{\sqrt{2k_{12}/J}}\right] \sin\left(\sqrt{\frac{2k_{12}}{J}}t\right)$$

(5.18)

The first term in both equations shows that both disks have a common angular displacement in the same direction that depends on the initial angular velocities of the disks. This is, of course, the rigid-body motion of the system associated with the "zero" natural frequency.

The second terms in the solution are of equal magnitude, but opposite sign, and represent periodic oscillations relative to the rigid-body motion.

From these solutions, we find that the disk–spring system rotates at a constant angular velocity while the disks oscillate back and forth about a mean angle, which is instantaneously located by the term $\frac{1}{2}[\dot{\theta}_1(0) + \dot{\theta}_2(0)]t$. These equations are general in that negative or positive initial velocities of any magnitude can be inserted.

Although the results were obtained for the specific case where $J_1 = J_2$, the conclusions drawn are applicable to the general situation when $J_1 \neq J_2$. In that case, the characteristic equation is

$$\begin{vmatrix} \dfrac{k_{12}}{J_1} - \omega^2 & \dfrac{-k_{12}}{J_1} \\[2ex] \dfrac{-k_{12}}{J_2} & \dfrac{k_{12}}{J_2} - \omega^2 \end{vmatrix} = 0$$

or

$$\left(\frac{k_{12}}{J_1} - \omega^2\right)\left(\frac{k_{12}}{J_2} - \omega^2\right) - \left(\frac{k_{12}}{J_1}\right)\left(\frac{k_{12}}{J_2}\right) = 0$$

(5.19)

which reduces to

$$\omega^2\left[\omega^2 - \left(\frac{k_{12}}{J_1} + \frac{k_{12}}{J_2}\right)\right] = 0$$

The roots of this equation are

$$\omega_n = 0 \text{ (two roots)}$$

$$\omega_n = \pm\sqrt{\frac{k_{12}}{J_1} + \frac{k_{12}}{J_2}}$$

(5.20)

The repeated zero roots indicate a rigid-body rotation. The natural frequency of oscillation, if any, will occur at the frequency

$$\omega_n = \sqrt{\frac{k_{12}}{J_1} + \frac{k_{12}}{J_2}}$$

From this exercise, we learn two things: zero roots signal the possibility of rigid-body motion, and the system's natural frequencies can be found by inspection of the decoupled differential equations. Uncoupling in this problem was achieved using a trick. More general methods for decoupling equations are available but not developed in this text.

5.6 CENTRIFUGAL-PENDULUM VIBRATION ABSORBER

In Section 5.2, we found that adding a mass–spring system with a fixed natural frequency to the frame of a machine acted as a vibration absorber, eliminating frame resonance at its natural frequency. Unfortunately, the two neighboring resonances that occur on either side of the canceled frequency limit the effectiveness of the device to a narrow range of frequencies near the original resonance. Absorbers of this type should be applied to machines that operate at very nearly constant speed.

For machines whose operating speed changes frequently, it would be nice to have an absorber whose natural frequency changes with the machine's speed, i.e., the frequency of the excitation. The pendulum absorber shown in Figure 5.10 satisfies this requirement for rotating shafts. Torsional systems, such as helicopter rotors, where the rotating shaft is excited by oscillations of the input torque, are examples of machines requiring such absorbers.

Figure 5.11 shows the absorber with a rotating coordinate system attached to a pendular member. The unit vectors $\mathbf{e}_r$ and $\mathbf{e}_\theta$ are fixed in this system. The XY coordinate system is fixed to ground. The velocity of the pendulum bob, mass m, is given by

$$\mathbf{V}_g = R\dot{\theta}\mathbf{e}_\theta + r(\dot{\phi} + \dot{\theta})(\cos\phi)\mathbf{e}_\theta - r(\dot{\phi} + \dot{\theta})(\sin\phi)\mathbf{e}_r$$

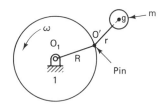

Figure 5.10 Centrifugal Pendulum Absorber

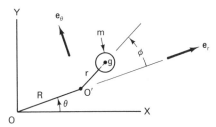

Figure 5.11 Coordinate System for Absorber

Neglecting the small potential energy variations due to changes in the elevation, the Lagrangian reduces to just the kinetic energy, i.e.,

$$L = T - V = T$$

$$= \tfrac{1}{2}m\{[R\dot{\theta} + r(\dot{\phi} + \dot{\theta})\cos\phi]^2 + [r(\dot{\phi} + \dot{\theta})\sin\phi]^2\} + \tfrac{1}{2}I_g(\dot{\theta} + \dot{\phi})^2$$

Since θ and $\dot{\theta}$ are the inputs, the generalized dependent coordinate is ϕ and the Lagrange equation of motion is

$$\frac{d}{dt}\left(\frac{\partial L}{\partial \dot{\phi}}\right) - \frac{\partial L}{\partial \phi} = Q_\phi$$

Because the pendulum undergoes a "free" motion, $Q_\phi = 0$.

By using the above Lagrangian, the Lagrange equation yields the differential equation of motion.

$$(I_g + mr^2)(\ddot{\phi} + \ddot{\theta}) + mRr(\ddot{\theta}\cos\phi + \dot{\theta}^2\sin\phi) = 0 \qquad (5.21)$$

Usually, $I_g \ll mr^2$, so that the equation can be simplified to

$$R\ddot{\theta}\cos\phi + R\dot{\theta}^2\sin\phi + r(\ddot{\phi} + \ddot{\theta}) = 0$$

For practical reasons, we would prefer not to have large variations in ϕ. If ϕ is small, the differential equation for ϕ linearizes to

$$\ddot{\phi} + \left(\frac{R}{r}\dot{\theta}^2\right)\phi = -\left(1 + \frac{R}{r}\right)\ddot{\theta} \qquad (5.22)$$

By inspection, the natural frequency of the pendulum absorber is

$$\omega_n = \dot{\theta}\sqrt{\frac{R}{r}} \qquad (5.23)$$

This is the desired result, i.e., the natural frequency of the absorber varies linearly with the shaft rotational speed $\dot{\theta}$.

The right-hand side of the differential equation shows that the absorber is excited by changes in the shaft speed. Ordinarily, the shaft displacement is composed of a steady rotation component $\dot{\theta}_o t$ and a superposed oscillatory portion, which will be taken to be $\theta_o \sin\omega t$, where ω is the frequency of the shaft oscillation. It is these oscillations that can lead to a fatigue failure of the shaft.

The shaft angular displacement is given by the equation:

$$\theta = \dot{\theta}_o t + \theta_o \sin\omega t$$

and the angular velocity and acceleration by

$$\dot{\theta} = \dot{\theta}_o + \theta_o\omega\cos\omega t$$

$$\ddot{\theta} = -\omega^2\theta_o\sin\omega t$$

For small amplitude excitations, $\theta_o\omega \ll \dot{\theta}_o$, with the result that $\dot{\theta} \cong \dot{\theta}_o$. The differential equation can now be written approximately as

$$\ddot{\phi} + \left(\frac{R}{r}\dot{\theta}_o^2\right)\phi = \left(1 + \frac{R}{r}\right)\theta_o\omega^2 \sin \omega t$$

Assuming a solution of the form

$$\phi = \phi_o \sin \omega t$$

yields the ratio

$$\frac{\theta_o}{\phi_o} = \frac{\left(\frac{R}{r}\right)\dot{\theta}_o^2 - \omega^2}{\left(\frac{R + r}{r}\right)\omega^2} \tag{5.24}$$

Recall that θ_o is the amplitude of the shaft oscillation, ϕ_o is the amplitude of the pendulus oscillation, ω is the frequency of the shaft oscillation, and $\dot{\theta}_o$ is the nominal shaft speed. Suppose this absorber was to be applied to a six-cylinder, four-stroke-cycle reciprocating engine crankshaft. There would be three power strokes each revolution of the shaft, so that the frequency of the shaft oscillation ω would be $3\dot{\theta}_o$. Inserting this into Equation (5.24) reduces it to

$$\frac{\theta_o}{\phi_o} = -\frac{\dfrac{R}{r} - 9}{9\left(\dfrac{R}{r} + 1\right)}$$

If the absorber is designed so that $R/r = 9$, then the amplitude of the shaft oscillation θ_o is zero regardless of the engine running speed $\dot{\theta}_o$.

The natural frequency of the pendulum in this case is

$$\omega_n = \dot{\theta}_o \sqrt{\frac{R}{r}} = 3\dot{\theta}_o$$

This is exactly the frequency of the shaft oscillation, ω. As with the translational absorber analyzed in Section 5.1, when this absorber's natural frequency is set equal to the frequency of the excitation, the amplitude of the excited component goes to zero. However, in this case, the absorber's natural frequency tracks the excitation frequency, which changes with the shaft speed, so that the absorber is effective at all shaft speeds.

5.7 WHIRLING OF A FLEXIBLE ROTOR SYMMETRICALLY MOUNTED ON RIGID SUPPORTS

In Section 5.4, we investigated the response of a flexibly mounted rigid machine to the unbalance of a rigid rotor. Unfortunately, rotors can rarely be treated as rigid members. Rotors such as we find in electric motors, turbines, compressors, pumps, etc., usually consist of relatively massive and stiff portions supported by much lighter-weight and flexible shafts. The rotor as a whole resembles a beam with concentrated or distributed loads, depending on the length of the more massive portion.

Figure 5.12 shows a thin disk mounted on a flexible shaft. The disk is equidistant from the rigid-shaft supports at either end. The center of mass of the disk–shaft assembly, g, is shown eccentric to the geometric center, c, of the straight disk–shaft rotor. The bow in the shaft, δ_{st}, shown in Figure 5.12, is the gravity-induced sag.

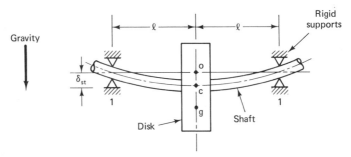

Figure 5.12 Disk Symmetrically Mounted on Flexible Shaft and Rigid Supports—Rest Configuration

Unbalance is inherent in all manufactured rotors. It can be decreased by balancing techniques (which will be discussed later in this chapter), but never be reduced to zero. The degree of acceptable unbalance depends on the particular application. Practically, some residual unbalance always remains no matter how carefully the balancing process is done. The remainder of this section will be devoted to an examination of the effect that this residual unbalance, e, has on the dynamics of a rotating rotor.

Figure 5.13 shows a side view of a whirling disk–shaft assembly. The displacement δ is the rotating bow of the shaft due to dynamic effects. The angular displacement of the plane of this bow is indicated by the angle ϕ. The bow distance δ is also the displacement of the center of the beamlike shaft, c. The bowed shaft acts like a spring, tending to restore the shaft center to point 0, which lies on the line of centers of the two bearings. (The rest bow, δ_{st}, supports the gravity load.)

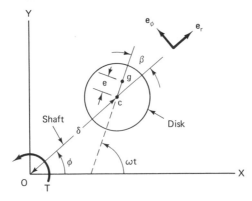

Figure 5.13 Side View of Whirling Rotor

The center of mass of the rotor is shown at g, which is a fixed distance from c. To keep the analysis general, the center of mass is not shown aligned with the shaft bow. The conditions that lead to alignment will be found as a consequence of the analysis, but we will not presume that alignment occurs a priori.

The angle β is the phase angle, which will be allowed to vary. Since the line joining c and g is the unbalance distance, which is fixed in the system, the angular velocity of the system, ω, is given by the sum of $\dot{\beta}$ and $\dot{\phi}$, i.e.,

$$\omega = \dot{\phi} + \dot{\beta} \tag{5.25}$$

This is the angular velocity of the rotor, which would be measured by a tachometer attached to the frame of a rigidly mounted machine.

The system shown in Figure 5.13 is analogous to the earth-sun solar system. Imagine that the disk is the earth, which is held to the sun by gravitational effects acting through the distance δ. The geometric center of the earth orbits (whirls) about the sun at 0 with a $\dot{\phi}$, such that it takes one year to complete one revolution. The angular velocity of the earth around its own axis, $\dot{\beta}$, causes the earth to make one revolution each day. The total angular velocity of the earth is the sum of these two angular velocities, the first absolute and the second relative.

Figure 5.13 shows a polar-coordinate system that rotates with the bowed shaft so that unit vector $\mathbf{e}_r$ always lies in the plane of the bow and $\mathbf{e}_\phi$ normal to the plane of the bow. The origin of this coordinate system is located by vector δ and has the velocity

$$\boldsymbol{\delta} = \dot{\delta}\mathbf{e}_r + \delta\dot{\phi}\mathbf{e}_\phi$$

The velocity of g relative to the $\mathbf{e}_r$, $\mathbf{e}_\phi$, and $\mathbf{e}_z$ moving coordinates is

$$\mathbf{v}_{r\phi} = e\dot{\beta}(\cos\beta\,\mathbf{e}_\phi - \sin\beta\,\mathbf{e}_r)$$

The velocity component due to the rotation of the coordinate system is

$$\boldsymbol{\dot{\phi}} \times \mathbf{e} = \dot{\phi}\mathbf{e}_z \times [e(\cos\beta)\mathbf{e}_r + e(\sin\beta)\mathbf{e}_\phi] = e\dot{\phi}[(\cos\beta)\mathbf{e}_\phi - (\sin\beta)\mathbf{e}_r]$$

The velocity of the center of mass is given by the standard formula

$$\mathbf{V}_g = \boldsymbol{\dot{\delta}} + \boldsymbol{\dot{\phi}} \times \mathbf{e} + \mathbf{v}_{r\phi} = \dot{\delta}\mathbf{e}_r + \delta\dot{\phi}\mathbf{e}_\phi + e(\dot{\phi} + \dot{\beta})[(\cos\beta)\mathbf{e}_\phi - (\sin\beta)\mathbf{e}_r]$$

Since $\omega = \dot{\phi} + \dot{\beta}$, this equation can be rewritten in the form

$$\mathbf{V}_g = (\dot{\delta} - e\omega\sin\beta)\mathbf{e}_r + (\delta\dot{\phi} + e\omega\cos\beta)\mathbf{e}_\phi$$

The total kinetic energy of the rotor is, therefore:

$$T = \tfrac{1}{2}m[(\dot{\delta} - e\omega\sin\beta)^2 + (\delta\dot{\phi} + e\omega\cos\beta)^2] + \tfrac{1}{2}J_g\omega^2$$

The spring constant of the shaft, k_r, can be determined from the load-deflection formula for a beam with a unit load applied at its midspan. If we neglect the effects of gravity, which are generally quite small, the potential energy due to the elastic deflection of the shaft is

$$V = \tfrac{1}{2}k_r\delta^2$$

Combining these into the Lagrangian yields

$$L = T - V = \tfrac{1}{2}m[(\dot{\delta} - e\omega \sin \beta)^2 + (\delta\dot{\phi} + e\omega \cos \beta)^2] + \tfrac{1}{2}J_g\omega^2 - \tfrac{1}{2}k_r\delta^2$$

There are three generalized coordinates in the Lagrangian, δ, ϕ, and β. The incremental work, dW, done on the rotor by the torque applied to the shaft, T, is

$$dW = T(d\phi + d\beta)$$

so that the generalized torques for the Lagrange equations are

$$Q_\phi = T \qquad Q_\beta = T$$

With the inputs given, the three Lagrange equations become

δ: $\quad \dfrac{d}{dt}\{m(\dot{\delta} - e\omega \sin \beta)\} - m(\delta\dot{\phi}^2 + e\omega\dot{\phi} \cos \beta) + k_r\delta = 0$

β: $\quad \dfrac{d}{dt}\left\{ me\left[\omega e\left(1 + \dfrac{J_g}{me^2}\right) - \dot{\delta} \sin \beta + \delta\dot{\phi} \cos \beta \right]\right\}$

$$+ me[\omega(\dot{\delta} \cos \beta + \delta\dot{\phi} \sin \beta)] = T$$

(5.26)

ϕ: $\quad \dfrac{d}{dt}\{m[\delta^2\dot{\phi} - e\dot{\delta} \sin \beta + \omega(e^2 + k^2) + e\delta(\omega + \dot{\phi}) \cos \beta] = T$

This set of equations is obviously very difficult to solve for the most general case. For machines that are run at a constant speed, except for startup and shutdown, we can make the simplification that $\dot{\omega} = 0$, so that $\ddot{\beta} = -\ddot{\phi}$.

The last three equations then reduce to

$$\ddot{\delta} + (\omega_n^2 - \dot{\phi}^2)\delta = e\omega^2 \cos \beta$$

$$(\delta\ddot{\phi} + 2\dot{\delta}\dot{\phi}) \cos \beta - (\ddot{\delta} - \delta\dot{\phi}^2) \sin \beta = T/em$$

$$\delta^2\ddot{\phi} + 2\delta\dot{\delta}\dot{\phi} + e[(\delta\ddot{\phi} + 2\dot{\delta}\dot{\phi}) \cos \beta - (\ddot{\delta} + \delta(\omega^2 - \dot{\phi}^2))e \sin \beta] = T/m$$

Eliminating the torque reduces these equations to

$$\ddot{\delta} + (\omega_n^2 - \dot{\phi}^2)\delta = e\omega^2 \cos \beta$$

$$\delta\ddot{\phi} + 2\dot{\delta}\dot{\phi} = e\omega^2 \sin \beta$$

(5.27)

where $\omega_n^2 = k_r/m$. The quantity ω_n is the natural frequency of the nonrotating rotor, considering the shaft mass to be negligible and the beam to act like a spring.

Since these equations are still too difficult to solve by ordinary methods, we will assume that β can remain constant, i.e., $\dot{\beta} = 0$. Then $\dot{\phi} = \omega$ and $\ddot{\phi} = \dot{\omega} = 0$. Now the shaft bow rotates at shaft speed while the phase angle remains locked at some angle β that remains constant. This motion is called synchronous precession, or whirl, for obvious reasons.

From the second equation, if $\ddot{\phi} = 0$ and $\dot{\phi} = \omega$, it reduces to

$$2\dot{\delta}\omega = e\omega^2 \sin \beta$$

from which we must conclude that $\dot{\delta}$ is constant since all the other terms in the equation are constant. If $\dot{\delta}$ is constant, then $\ddot{\delta} = 0$, which means that the first equation reduces to

$$(\omega_n^2 - \omega^2)\delta = e\omega^2 \cos \beta$$

or

$$\delta = \left(\frac{e\omega^2}{\omega_n^2 - \omega^2}\right) \cos \beta$$

Since the right-hand side of this equation is also constant, δ must be constant. Returning to the second equation, with δ = constant, we are left with

$$0 = e\omega^2 \sin \beta$$

which means that the only allowable constant values of β are $0°$ and $180°$.

Before proceeding further, let us review the results of our assumption that β could remain constant. Inspection of the equations has shown that for the hypothesis to be true, the shaft bow must rotate (precess) at shaft speed, the shaft displacement δ must remain constant, and the phase angle β must be either $0°$ or $180°$. The equation for the shaft displacement under these conditions is

$$\delta = \left(\frac{e\omega^2}{\omega_n^2 - \omega^2}\right) \cos \beta \tag{5.28}$$

The only remaining question that must be satisfactorily answered if we are to justify the original assumption is "which of the phase angles is the correct one?" The answer to this question leads to a very startling discovery.

Suppose that the shaft speed is less than the system's natural frequency. Since δ must be positive, β must be zero, so that $\cos \beta = +1.0$. Geometrically, this means that the shaft and mass centers whirl in perfect circles about 0. The radius of the mass center's orbit will be larger than the radius of the shaft center's orbit δ. Intuitively, this conclusion is quite acceptable because we expect to find that the "heavy side" of the system is thrown to the "outside" by centrifugal effects.

If, on the other hand, the shaft speed ω is larger than the rotor's natural frequency ω_n, then δ will be negative, unless $\beta = 180°$. If $\beta = 180°$, then $\cos \beta = -1.0$, which means that the mass center orbits at a radius that is less than the orbit radius of the geometric center. Now the "heavy side" is "inside," which is contrary to our intuition. The switch in radii that apparently occurs as the shaft speed passes through the rotor's natural frequency is called an "inversion," since g and c somehow change their relative positions. Figure 5.14 illustrates the two states of synchronous whirl just discovered.

The explanation for the inversion is quite simple. At higher speeds, the centrifugal effect becomes quite large. To balance this outward force, the shaft must deflect more. It achieves this larger deflection by moving c from inside of g to outside of g as the rotor's speed passes through its natural frequency. Since the shaft

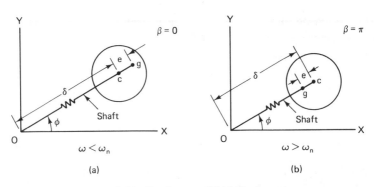

Figure 5.14   Synchronous Whirl Configuration

deflection tends to be quite large when $\omega = \omega_n$, it is important to pass through the inversion speed rather quickly, relying on inherent damping to limit the amplitude of δ at resonance.

The results obtained before represent the steady-state end-point behavior of the system. The time-dependent analysis of the actual inversion process during which $\dot{\beta} \neq 0$ is beyond the scope of this text. Synchronous whirl and inversion at the natural frequency are well-known observed phenomena and not just a hypothesis.

As with many phenomena, the observation of synchronous whirl probably preceded the analytical explanation just developed. This "synchronous-whirl" solution is undoubtedly one of many possible solutions of the nonlinear, coupled differential equations of motion. Fortunately, we have been able to provide a simple explanation for it. Since we have not obtained a general solution of the differential equations, there is no guarantee that other more complex motions cannot occur under certain circumstances. Obviously, we have only scratched the surface of a very complicated subject.

As a matter of practical interest, the natural frequency of the rotor can be determined by measuring or computing the gravity-induced sag of the shaft, δ_{st}. When at rest, the weight of the shaft is supported by the static bend of the shaft, so that equilibrium requires

$$W = k_r \delta_{st}$$

or

$$\frac{g}{\delta_{st}} = \frac{k_r}{W/g} = \frac{k_r}{m}$$

Then

$$\omega_n = \sqrt{\frac{k_r}{m}} = \sqrt{\frac{g}{\delta_{st}}}$$

We shall see in the next section that this result is also a special case of an approximate method for finding the lowest natural frequency of a multidisk rotor.

5.8 MULTIDISK ROTORS

Many rotors consist of several disklike elements attached to a shaft that is not symmetrically supported on rigid bearings. The results obtained in the previous section are obviously not applicable to such configurations. Even a single-disk rotor, where the disk and supporting-shaft diameters are not symmetric relative to the supports, cannot be treated by the methods developed so far.

In this section, we will develop a more general (matrix) method for finding the natural frequencies of multidisk rotors. We will also obtain two approximate methods for finding the lowest natural frequency of a system, one attributed to Rayleigh and the other to Dunkerley.

The natural-frequency formulas given at the end of the last section can also be obtained by assuming that the maximum kinetic energy of the vibrating rotor is equal to the maximum of the elastically stored energy of the rotor. For a conservative system vibrating at its natural frequency, the sum of the kinetic and potential energies is constant. At the extremes of the vibration, the rotor's kinetic energy is zero and its stored energy is a maximum. At the midpoint of its vibration, the kinetic energy is maximum and the stored energy of the straight shaft is zero. Since both maxima are equal to the same constant, they are also equal to each other.

For the rotor examined in the last section, the maximum kinetic energy is

$$T_m = \frac{1}{2} \frac{W}{g} (\omega \delta)^2$$

while the maximum stored energy is

$$V_m = \tfrac{1}{2} k_r (\delta)^2$$

Equating these two maxima,

$$\frac{1}{2} \frac{W}{g} (\omega \delta)^2 = \frac{1}{2} k_r (\delta)^2$$

canceling the δ's, and solving for ω yields

$$\omega = \sqrt{\frac{k_r g}{W}} \tag{5.29}$$

This is the exact equation for the natural frequency of a single disk symmetrically supported on a flexible shaft. Notice that it is not necessary to know the stiffness of the disk in this case. Since $W/k_r = \delta_{st}$, the equation can be reduced to

$$\omega_n = \sqrt{\frac{g}{\delta_{st}}}$$

For n disks on a common shaft, Rayleigh proposed that the kinetic energy be approximated by extending the single-disk equation to the summation:

$$T_m = \frac{\omega_n^2}{2g} (W_1 \delta_1^2 + W_2 \delta_2^2 + \cdots + W_n \delta_n^2) \tag{5.30}$$

where δ_1, δ_2, etc. are the *static* displacements under each of the disks W_1, W_2, . . . , W_n, as shown in Figure 5.15.

You will recall that the spring constant was eliminated from the natural-frequency formula (5.29) by noting that $W = k_r \delta_{st}$. Multiplying both sides of this equation by $\frac{1}{2} \delta_{st}$ yields the energy relation:

$$\tfrac{1}{2} W \delta_{st} = \tfrac{1}{2} k_r \delta_{st}^2$$

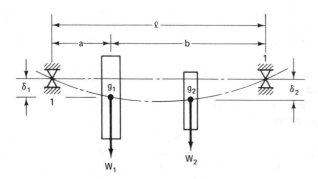

Figure 5.15 Rotor with Multiple Disks

Rayleigh also proposed that this expression be extended to sum the stored elastic energy, so that

$$V_m = \tfrac{1}{2}(W_1 \delta_1 + W_2 \delta_2 + \cdots + W_n \delta_n) \tag{5.31}$$

Equating these two expressions and solving for ω_n, he obtained

$$\omega_n^2 = g \frac{W_1 \delta_1 + W_2 \delta_2 + \cdots + W_n \delta_n}{W_1 \delta_1^2 + W_2 \delta_2^2 + \cdots + W_n \delta_n^2} \tag{5.32}$$

For a single disk, this equation reduces to the exact equation for ω_n. It is approximate for multiple disks because of the assumption that the dynamic displacements at each disk are equal to the static displacements. That is not exactly true.

Because the d'Alembert inertial forces will cause the actual displacements to be larger than the static displacements, the energies in both the denominator and numerator will be underestimated by the Rayleigh formulation. However, the error in the underestimate will be larger in the denominator since it involves the squares of the approximated displacements. As a consequence, the formula overestimates the lowest natural frequency.

To obtain the total displacement inputs δ_1 and δ_2 of Figure 5.15, we could solve the two separate static-beam displacement problems shown in Figure 5.16. In Figure 5.16(a), where W_2 is absent, δ_{11} is the displacement at point 1 due to the application of W_1 at point 1. The displacement δ_{21} is the displacement at point 2 when W_1 is applied at point 1. The same interpretation applies to the displacements shown in Figure 5.16(b), where W_1 is absent. The total displacements required for the Rayleigh estimate are found by superposition, so that

$$\delta_1 = \delta_{11} + \delta_{12}$$

$$\delta_2 = \delta_{21} + \delta_{22}$$

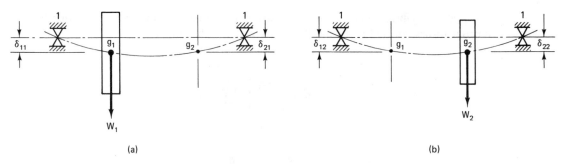

(a)   (b)

Figure 5.16   Beam Displacements

Each of the contributing deflections on the right-hand side of the formulas will be proportional to the corresponding load that causes it. For example,

$$\delta_{11} = a_{11} W_1 \qquad \delta_{21} = a_{21} W_1$$

$$\delta_{12} = a_{12} W_2 \qquad \delta_{22} = a_{22} W_2$$

where the coefficients a_{ij} are obtained from the standard strength-of-materials beam-deflection formulas.[†] The total deflections and the loads, therefore, can be related by a "flexibility" matrix $[A]$, i.e.,

$$\left\{ \begin{matrix} \delta_1 \\ \delta_2 \end{matrix} \right\} = [A] \left\{ \begin{matrix} W_1 \\ W_2 \end{matrix} \right\}$$

where

$$[A] = \begin{bmatrix} a_{11} & a_{12} \\ a_{21} & a_{22} \end{bmatrix} \tag{5.33}$$

The elements of this matrix are called the flexibility-influence coefficients.

The inverse of the *flexibility matrix* is the "stiffness" matrix $[k]$, which relates displacements to loads through the matrix equation:

$$\left\{ \begin{matrix} W_1 \\ W_2 \end{matrix} \right\} = [k] \left\{ \begin{matrix} \delta_1 \\ \delta_2 \end{matrix} \right\}$$

Once the coefficients of the flexibility matrix are determined, the displacements δ required by the Raleigh approximation are readily calculated. However, it is not necessary to settle just for an approximate answer! The flexibility matrix permits us to formulate the characteristic equation for *all* the natural frequencies, as we will now demonstrate.

[†] In Figure 5.16(a),

$$a_{11} = \frac{l_1^2 (l - l_1)^2}{3EIl}$$

$$a_{21} = \frac{(l - l_1)^2 l_2^2}{6EI} [l_2^2 + (l - l_1)^2 - l^2]$$

For the vibrating rotor, the loads are not the disk weights, but the d'Alembert inertial forces. In that case, the flexibility equation becomes

$$\left\{\begin{matrix} \delta_1 \\ \delta_2 \end{matrix}\right\} = [A]\left\{\begin{matrix} -\dfrac{W_1}{g}\ddot{\delta}_1 \\[2mm] -\dfrac{W_2}{g}\ddot{\delta}_2 \end{matrix}\right\}$$

For *harmonic motion,*

$$\ddot{\delta}_1 = -\omega^2\delta_1$$
$$\ddot{\delta}_2 = -\omega^2\delta_2$$

so that the flexibility equation becomes

$$\frac{1}{\omega^2}\left\{\begin{matrix} \delta_1 \\ \delta_2 \end{matrix}\right\} = \begin{bmatrix} a_{11}\dfrac{W_1}{g} & a_{12}\dfrac{W_2}{g} \\[2mm] a_{21}\dfrac{W_1}{g} & a_{22}\dfrac{W_2}{g} \end{bmatrix}\left\{\begin{matrix} \delta_1 \\ \delta_2 \end{matrix}\right\}$$

or

$$\begin{bmatrix} a_{11}\dfrac{W_1}{g} - \dfrac{1}{\omega^2} & a_{12}\dfrac{W_2}{g} \\[2mm] a_{21}\dfrac{W_1}{g} & a_{22}\dfrac{W_2}{g} - \dfrac{1}{\omega^2} \end{bmatrix}\left\{\begin{matrix} \delta_1 \\ \delta_2 \end{matrix}\right\} = \left\{\begin{matrix} 0 \\ 0 \end{matrix}\right\}$$

The characteristic equation for this system is the determinant of this matrix set equal to zero, i.e.,

$$\begin{vmatrix} a_{11}\dfrac{W_1}{g} - \dfrac{1}{\omega^2} & a_{12}\dfrac{W_2}{g} \\[2mm] a_{21}\dfrac{W_1}{g} & a_{22}\dfrac{W_2}{g} - \dfrac{1}{\omega^2} \end{vmatrix} = 0 \qquad (5.34)$$

The roots of this equation are the eigenvalues, or the natural frequencies, of the system. Since there are two degrees of freedom, two natural frequencies, ω_{n1} and ω_{n2}, can be determined.

The flexibility matrix is used extensively in the analysis of structures. With complex structures, it is usually not possible or practical to compute the elements of this matrix, in which case, they are obtained experimentally by applying loads at the various points of interest and measuring the resulting displacements there, and at the other points.

For a multidisk system with n masses, there would be n natural frequencies. For systems where we are only interested in the lowest of these, Dunkerley has provided us with an approximate formula for its determination. You will recall that Rayleigh's approximation provides an upper bound on the lowest natural frequency.

We will find that Dunkerley provides the lower bound on this frequency. Using both approximations, we can bracket the exact lowest natural frequency.

If the determinant form of the characteristic equation is expanded, it becomes

$$\left(\frac{1}{\omega^2}\right)^2 - \left(a_{11}\frac{W_1}{g} + a_{22}\frac{W_2}{g}\right)\left(\frac{1}{\omega^2}\right) + (a_{11}a_{22} - a_{12}a_{21})\frac{W_1}{g}\frac{W_2}{g} = 0$$

From algebra, we know that the coefficient of the second term is equal to the sum of the roots of the equation, i.e.,

$$\frac{1}{\omega_{n1}^2} + \frac{1}{\omega_{n2}^2} = a_{11}\frac{W_1}{g} + a_{22}\frac{W_2}{g}$$

Dunkerley observed that $\omega_{n2} > \omega_{n1}$, so that if $1/\omega_{n2}^2$ is neglected in the last equation,

$$\frac{1}{\omega_{n1}^2} < a_{11}\frac{W_1}{g} + a_{22}\frac{W_2}{g}$$

or, conversely,

$$\omega_{n1}^2 > \frac{1}{a_{11}\dfrac{W_1}{g} + a_{22}\dfrac{W_2}{g}}$$

The latter equation provides a lower bound on ω_{n1} to complement the upper bound of Rayleigh's approximation.

The utility of this approximation becomes evident when we interpret the meaning of the terms on the right-hand side of the inequality. As noted earlier, $\delta_{11} = a_{11}W_1$ and $\delta_{22} = a_{22}W_2$, so that

$$a_{11}\frac{W_1}{g} = \frac{\delta_{11}}{g} \qquad a_{22}\frac{W_2}{g} = \frac{\delta_{22}}{g}$$

Since the displacements in these equations are the displacements at the loads when only they are present, δ_{11}/g is the reciprocal of the square of the natural frequency of the rotor when only disk (1) is attached. Similarly, δ_{22}/g is the reciprocal of the square of the rotor's natural frequency when only disk (2) is attached. Designating these as

$$\omega_{11}^2 = \frac{g}{\delta_{11}} \qquad \omega_{22}^2 = \frac{g}{\delta_{22}}$$

then

$$\frac{1}{\omega_{n1}^2} < \frac{1}{\omega_{11}^2} + \frac{1}{\omega_{22}^2} \tag{5.35}$$

Note that Dunkerley's approximation uses only the displacements δ_{11} and δ_{22}, whereas these must be combined with δ_{12} and δ_{21} to obtain Rayleigh's approximation.

Dunkerley's approximation is reminiscent of the results obtained in Section 5.2, where the natural frequencies of the whole system were found to be dependent upon the natural frequencies of component mass–spring systems.

From a practical point of view, Dunkerley's approximation is probably the most useful since it requires fewer coefficients and underestimates the lowest natural frequency. The Rayleigh approximation requires the determination of all the flexibility coefficients found in the characteristic determinant. With these in hand, one might consider, instead, calculating the "exact" natural frequencies from the characteristic equation. The word exact is in quotation marks because the flexibility-matrix formulation does not account for gyroscopic effects, a topic we will consider in the next section.

5.9 GYROSCOPIC EFFECTS

Gyroscopic effects occur when the geometric axes of the rotor disks change their direction with time. The axis of the symmetrically mounted single disk shown in Figure 5.12 whirls about the bearing axis without changing the direction of its axis. On the other hand, when two disks are mounted on a common shaft, as shown in Figure 5.16(a), their axes do change directions as the disks whirl about the bearing's line of centers. The treatment of multidisk rotors in the last section took no account of how this motion influences the rotor's natural frequencies.

In this section we will determine how gyroscopic effects influence the natural frequency of an overhung disk. The trend established for this configuration is the same for all rotor configurations. Figure 5.17 shows the end portion of a rotor that consists of an overhung disk on a cantilevered, flexible shaft. To simplify the problem, the rotor will be assumed to be perfectly balanced. The shaft, however, will be assumed to be slightly bent as a result of manufacturing imperfections.

As before, $\dot{\phi}$ is the whirl angular velocity of the plane of the bent shaft, and $\dot{\beta}$ is the phase angular velocity of the disk relative to that plane. The direction of the

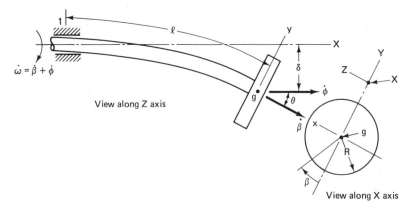

Figure 5.17   Overhung Disk on Flexible Shaft

vectors corresponding to these velocities are shown in Figure 5.17. The components of the total angular velocity of the disk normal to the disk and in the plane of the disk are, respectively, given by

$$\omega_x = \dot{\beta} + \dot{\phi} \cos \theta$$

and

$$\omega_y = \dot{\phi} \sin \theta$$

The velocity of the mass center of the disk tangent to its orbit is

$$V_g = \delta \dot{\phi}$$

There are two moments of inertia that must be considered here. One is about the disk axis, J_p, and the other about the disk diameter, J_d.[†] The total kinetic energy of the disk, taking into account the two components of the rotational kinetic energy, is

$$T = \tfrac{1}{2} m (\delta \dot{\phi})^2 + \tfrac{1}{2} J_p (\dot{\beta} + \dot{\phi} \cos \theta)^2 + \tfrac{1}{2} J_d (\dot{\phi} \sin \theta)^2$$

In this problem, it will be convenient to introduce the effects of the shaft stiffness as applied generalized forces rather than as contributors to the total potential energy V. This is an option we have frequently exercised when setting up the Lagrange equations, and the only way we introduced damping into them. Neglecting the effects of gravity, the potential energy is zero.

The generalized forces acting on the disk are denoted as

$Q_\delta = F =$ positive force in the positive δ direction

$Q_\theta = M =$ positive moment in the positive θ direction

The two Lagrange equations that apply to the system, i.e., the disk, are

$$\frac{d}{dt}\left(\frac{\partial L}{\partial \dot{\delta}}\right) - \frac{\partial L}{\partial \delta} = Q_\delta$$

and

$$\frac{d}{dt}\left(\frac{\partial L}{\partial \dot{\theta}}\right) - \frac{\partial L}{\partial \theta} = Q_\theta$$

where $L = T$.

From the first of these dynamic equations, we obtain the result

$$-m\delta (\dot{\phi})^2 = F \tag{5.36}$$

This equation says that the force on the disk is opposite to the outward centrifugal force. The force on the shaft is also outward, tending to increase θ, as we would expect.

[†] For a thin disk, $J_p = \tfrac{1}{2} m R^2$ and $J_d = \tfrac{1}{4} m R^2$

The second Lagrange equation yields

$$(J_p - J_d)(\dot{\phi})^2 \sin\theta \cos\theta + J_p \dot{\beta}\dot{\phi} \sin\theta = M \qquad (5.37)$$

For small shaft-bending angles θ, which is all we would tolerate in a well-constructed rotor, this equation for the gyroscopically induced couple reduces to

$$(J_p - J_d)\dot{\phi}^2\theta + J_p \dot{\beta}\dot{\phi}\theta = M$$

For simplicity, we will denote $J_p - J_d$ by J_{pd}. In the case of a thin disk,

$$J_{pd} = \tfrac{1}{4} mR^2$$

Since J_{pd} is positive, M is positive when $\dot{\beta} = 0$, which means that the couple that the end of the shaft exerts on the disk is also positive. The couple that the disk exerts on the shaft is, therefore, negative, an effect that tends to reduce θ and *straighten* the shaft. This is, of course, opposite to the centrifugal-force effect. The result is that there are two competing effects acting on the shaft's end that try to cancel each other.

The flexibility matrix for a cantilever beam with a force F and a couple M applied to its end can be obtained directly from strength-of-materials analysis, so that

$$\begin{Bmatrix} \delta \\ \theta \end{Bmatrix} = \frac{1}{EI}\begin{bmatrix} l^3/3 & l^2/2 \\ l^2/2 & l \end{bmatrix}\begin{Bmatrix} F \\ M \end{Bmatrix} \qquad (5.38)$$

Note that the off-diagonal coupling coefficients of the flexibility matrix are equal. This will always be the case with linear elastic elements.

Although the inherent unbalance of the rotor has been neglected, our earlier analysis of rotor systems has shown that whirl occurs at the synchronous speed, so that $\omega = \dot{\phi}$ *and* $\dot{\beta} = 0$. Inserting this simplification and the dynamic equations for the force F and the couple M into the flexibility equations, they become

$$\left(1 - \frac{ml^3\omega^2}{3EI}\right)\delta + \left(\frac{J_{pd}l^2\omega^2}{2EI}\right)\theta = 0$$

$$\left(\frac{ml^2\omega^2}{2EI}\right)\delta - \left(1 + \frac{J_{pd}l\omega^2}{EI}\right)\theta = 0$$

The determinant of the matrix of the coefficients of δ and θ set equal to zero is the characteristic equation for this system, i.e.,

$$\left(1 - \frac{ml^3\omega^2}{3EI}\right)\left(1 + \frac{J_{pd}l\omega^2}{EI}\right) + \left(\frac{ml^2\omega^2}{2EI}\right)\left(\frac{J_{pd}l^2\omega^2}{2EI}\right) = 0$$

The natural frequency of a vibrating, nonrotating cantilever with a concentrated load applied to its end is given by

$$\omega_n = \sqrt{\frac{3EI}{ml^3}}$$

The characteristic equation for the critical frequency of the whirling rotor can be written in terms of this frequency as follows:

$$\left(\frac{\omega}{\omega_n}\right)^4 + 4\left(\frac{1}{\alpha} - 1\right)\left(\frac{\omega}{\omega_n}\right)^2 - \frac{4}{\alpha} = 0 \tag{5.39}$$

where

$$\alpha = \frac{3J_{pd}}{ml^2} \tag{5.40}$$

Figure 5.18(a) shows the ratio of the critical frequency, ω_{cr}, of the rotor with gyroscopic effects to the natural frequency of the vibrating shaft, ω_n. Since the gy-

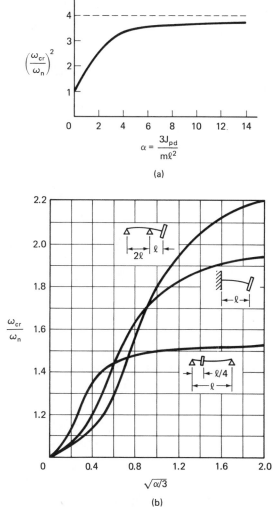

(a)

(b)

Figure 5.18 (a) Effect of Gyroscopic Couple on the Natural Frequency of Overhung Disk, (b) Comparison of Gyroscopic Effects

roscopic couple acts to stiffen the shaft against the whirl-induced centrifugal force, the critical speed of the rotor at which large-amplitude displacements occur is higher than the natural frequency of the rotor.

The implications of this result are multifold. First, gyroscopic effects could be beneficial if they raise the critical frequency above the shaft (excitation) frequency. Second, frequencies obtained from the "exact" matrix model developed in the previous section would err on the low side of the critical frequencies. Third, and probably most practical, it is common practice to determine a rotor's natural frequency by striking it with a hammer and measuring the frequency of the free vibration. In the case studied here, the frequency obtained by that method would be ω_n. However, when the shaft rotates, its critical frequency, ω_{cr}, will be larger than ω_n. Thus, this technique, like the Dunkerley approximation, yields a lower bound on the exciting frequency which causes resonance.

The effect of the gyroscopic couple on other single-disk rotor configurations is also shown in Figure 5.18(b). In each case, the rotor's critical frequency is larger than its natural frequency with an applied concentrated load. Based on these results, we can safely conclude that gyroscopic effects cause a rotor's critical, or resonant, frequency to be larger than its static natural frequency.

5.10 DYNAMICS OF A RIGID ROTOR WITH ASYMMETRIC-SUPPORT STIFFNESS

Figure 5.19 shows a rigid shaft with a single unbalanced disk attached at its axial midpoint. The shaft is supported at its ends by unequal horizontal and vertical springs. Unequal support stiffness can arise as a result of the bearing characteristics or because the portion of the machine supporting the bearing has a different stiffness in the horizontal and vertical directions. If the machine frame is pictured as a simple vertical cantilever beam fixed to ground at its lower end, it is not difficult to imagine that the horizontal (bending) stiffness might be noticeably different from its vertical (tensile or compressive) stiffness.

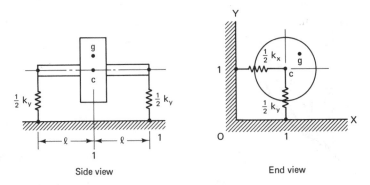

Figure 5.19   Rigid Rotor on Unequal Supports

Since these stiffnesses are attached to a fixed reference frame, the location of the center of mass will be described in terms of their horizontal and vertical components relative to the fixed frame. Figure 5.20 shows the appropriate geometry.

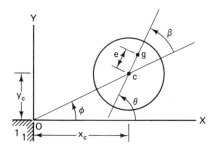

Figure 5.20   Coordinate System for Rotor

The coordinates of the center of mass are given by

$$x_g = x_c + e \cos (\phi + \beta)$$

$$y_g = y_c + e \sin (\phi + \beta)$$

where e is the fixed unbalance distance. In these equations, the angle ϕ locates the plane of the bow of the shaft. The angle θ locates the plane passing through the geometric center of the disk and the center of mass. The unbalance distance e lies in this plane. The angle between these two planes is the phase angle β. The angles are related by the expression

$$\theta = \phi + \beta$$

and the angular velocities by

$$\dot{\theta} = \dot{\phi} + \dot{\beta}$$

Since the unbalance distance, e, is fixed on the disk, θ is the total angular velocity of the rotor observed from the fixed reference frame XY. The angular velocity $\dot{\phi}$ is the velocity at which the disk center whirls about the bearing center line O.

If the mass-center displacements are written in the form

$$x_g = x_c + e \cos \theta$$

$$y_g = y_c + e \sin \theta$$

the generalized coordinates become x_c, y_c, and θ. The kinetic energy for the system is

$$T = \tfrac{1}{2}m[(\dot{x}_c - e\dot{\theta} \sin \theta)^2 + (\dot{y}_c + e\dot{\theta} \cos \theta)^2] + \tfrac{1}{2}I_g(\dot{\theta})^2$$

The potential energy, neglecting the small changes in the elevation of the mass center, is

$$V = \tfrac{1}{2}k_x(x_c)^2 + \tfrac{1}{2}k_y(y_c)^2$$

The Lagrangian is then

$$L = \tfrac{1}{2}m[(\dot{x}_c - e\dot{\theta}\sin\theta)^2 + (\dot{y}_c + e\dot{\theta}\cos\theta)^2] + \tfrac{1}{2}I_g(\dot{\theta})^2 - \tfrac{1}{2}k_x(x_c)^2 - \tfrac{1}{2}k_y(y_c)^2$$

To simplify the analysis, we will again examine the commonly encountered condition where the shaft speed $\omega = \dot{\theta}$ is constant, so that $\theta = \omega t$.

Three Lagrange equations can be written for this system. Two will be examined here; the third is the subject of an exercise in the problems at the end of the chapter. The equation

$$\frac{d}{dt}\left(\frac{\partial L}{\partial \dot{x}_c}\right) - \frac{\partial L}{\partial x_c} = Q_{xc} \qquad Q_{xc} = 0$$

becomes

$$m\ddot{x}_c + k_x x_c = me\omega^2 \cos\omega t \tag{5.41a}$$

The equation

$$\frac{d}{dt}\left(\frac{\partial L}{\partial \dot{y}_c}\right) - \frac{\partial L}{\partial y_c} = Q_{yc} = 0$$

yields

$$m\ddot{y}_c + k_y y_c = me\omega^2 \sin\omega t \tag{5.41b}$$

Since these two differential equations are uncoupled, the natural frequencies of the system can be determined by inspection. They are

$$\omega_{n1} = \sqrt{\frac{k_x}{m}} \qquad \omega_{n2} = \sqrt{\frac{k_y}{m}}$$

The steady-state solutions of the differential equations (assuming that the small amount of inherent damping present will cause the starting transient to decay to zero) are of the form

$$x_c = A\cos\omega t \qquad y_c = B\sin\omega t$$

Solving for the amplitudes A and B yields the solutions

$$x_c = \left[\frac{e\left(\dfrac{\omega}{\omega_{n1}}\right)^2}{1 - \left(\dfrac{\omega}{\omega_{n1}}\right)^2}\right]\cos\omega t = A\cos\omega t$$

$$y_c = \left[\frac{e\left(\dfrac{\omega}{\omega_{n2}}\right)^2}{1 - \left(\dfrac{\omega}{\omega_{n2}}\right)^2}\right]\sin\omega t = B\sin\omega t \tag{5.42}$$

From these, the trajectory of the center of the disk can be obtained as

$$(\sin \omega t)^2 + (\cos \omega t)^2 = \left(\frac{x_c}{A}\right)^2 + \left(\frac{y_c}{B}\right)^2 = 1$$

If A and B are unequal, the orbit of c, i.e., the whirl trajectory, is an ellipse. This ellipse degenerates into a circle when $A = B$ which is the whirl case studied earlier in Section 5.7.

From Figure 5.20, we see that

$$\tan \phi = \frac{y_c}{x_c} = \frac{B}{A} \tan \theta$$

The velocity of the shaft whirl is, therefore,

$$\dot{\phi} = \frac{d}{dt}\left[\tan^{-1}\left(\frac{B}{A}\tan\theta\right)\right] = \frac{AB\omega}{(A\cos\omega t)^2 + (B\cos\omega t)^2}$$

From this we find that when $A \neq B$, the whirl velocity is not constant, even when the shaft speed ω is constant. Only when $A = B$ does the whirl speed synchronize exactly with the shaft speed, so that $\omega = \dot{\phi}$. Of course, when k_x and k_y are very nearly equal, it might be very hard to detect this difference, which must result in a varying phase angle, β.

The trajectory of the center of mass is also elliptical. The coordinates of g can now be written:

$$x_g = (A + e) \cos \omega t$$

$$y_g = (B + e) \sin \omega t$$

By following the same procedure, the elliptical orbit of the center of mass is found from

$$\left(\frac{x_g}{A + e}\right)^2 + \left(\frac{y_g}{B + e}\right)^2 = 1$$

From the formulas for x_c and y_c, the amplitudes of the trajectory are

$$A = \frac{e(\omega/\omega_{n1})^2}{1 - (\omega/\omega_{n1})^2}$$

$$B = \frac{e(\omega/\omega_{n2})^2}{1 - (\omega/\omega_{n2})^2}$$

These major and minor radii tend to become very large when the shaft speed, ω, approaches either of the two natural frequencies, ω_{n1} or ω_{n2}. Resonant conditions like these, of course, should be avoided during the machine's operation.

The two critical rotor frequencies divide the operating-speed range into three separate regimes. When examining the system's performance, it will be assumed that ω_{n1} is greater than ω_{n2}, which means that the supports are stiffer in the horizontal direction than in the vertical direction. As a consequence, the major diameters of the orbits in the x direction will be smaller than the major diameters in the y direction.

When $\omega < \omega_{n2} < \omega_{n1}$, both A and B are positive $(A < B)$, so that the whirl $(\dot{\phi})$ is in the same direction as ω. Shaft whirl is then in the forward direction with the mass center's orbit "outside," as illustrated in Figure 5.21.

At the other extreme, when $\omega_{n2} < \omega_{n1} < \omega$, both A and B are negative, with $|A| < |B|$. The orbit of the center of mass is inside the orbit of the geometric center, so that the mass hangs "in" rather than "out." This is analogous to the inversion obtained earlier. Figure 5.22 shows the whirl objects, which are still both forward, as indicated by the equation relating ω and $\dot{\phi}$.

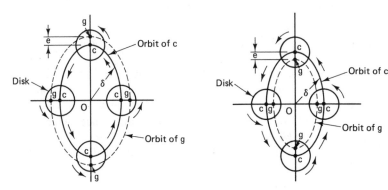

Figure 5.21   Forward Whirl, $\omega < \omega_{n_2} < \omega_{n_1}$

Figure 5.22   Forward Whirl, $\omega_{n_2} < \omega_{n_1} < \omega$

Lastly, when $\omega_{n2} < \omega < \omega_{n1}$, A is positive, but B is negative, so the whirl $\dot{\phi}$ is opposite to the shaft speed ω, i.e., backward, or retrograde, whirl occurs. We also find that the two orbits cross each other so that the mass center alternates from "inside" to "outside" twice each complete orbit, as shown in Figure 5.23. This crossing causes β to change during each orbit, giving rise to the cyclic stressing of the shaft twice each revolution. Cyclic stressing can ultimately lead to fatigue if the stress levels are too high.

Unless the horizontal and vertical stiffness of the supports differ greatly, the two critical speeds will not be widely separated. In that case, it would be unwise to even consider operating between these two criticals for fear that slight speed changes might cause the shaft to slip into one of the resonances. That consideration, and the possibility of fatigue, practically rules out trying to operate the rotor between its critical speeds.

The important point here is that the perfectly circular orbits predicted by the earlier model of a flexible shaft do not occur when the supporting system does not have a uniform stiffness in all directions.

When the stiffness of the shaft and its supports are comparable, an equivalent stiffness must be used. Figure 5.24 models an elastic shaft and support system. At

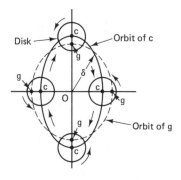

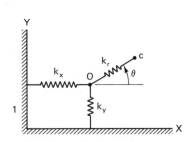

Figure 5.23 Backward Whirl, $\omega_{n_2} < \omega < \omega_{n_1}$

Figure 5.24 Flexible Shaft—Support System

point O, the horizontal and vertical spring force components are equal, i.e., in series. The horizontal force exerted by the shaft is $k_r \delta \cos \phi$. The quantity $\delta \cos \phi$ is the horizontal component of the shaft bow, so that the equivalent horizontal stiffness is given by the series-spring equation:

$$\frac{1}{(K_x)_{\text{eq}}} = \frac{1}{k_x} + \frac{1}{k_r} \tag{5.43a}$$

By a similar argument, the equivalent vertical stiffness, $(K_y)_{\text{eq}}$, is given by

$$\frac{1}{(K_y)_{\text{eq}}} = \frac{1}{k_y} + \frac{1}{k_r} \tag{5.43b}$$

The equivalent stiffnesses are less than any of the component stiffnesses used to compute them. Each stiffness added between the rotor and ground tends to lower the resonant frequency of the system. A reasonable approximation of the natural frequencies requires that all the supporting stiffnesses be included in the equivalent stiffnesses, even if some of them have to be estimated.

5.11 EFFECT OF BEARINGS ON ROTOR DYNAMICS

Bearing systems fall into two general categories, hydrodynamic and rolling element. Rolling-element bearings use balls or rollers to support the shaft load with low frictional resistance. Lubrication is achieved by packing grease in the space between the rollers or spraying a mist of oil into the bearing. Although the supporting structure formed by the rolling elements is discontinuous and moving, the bearing as a whole may still be treated as though it were a solid elastic springlike element. Spring constants for rolling-element bearings usually fall into the range of one to two million pounds per inch in the direction of the load application. They act in series with the shaft and support stiffnesses and combine with them according to the reciprocal summation equation.

Fluid film bearings are quite another matter. Unfortunately, they cannot be treated as a simple direct spring. Although the fluid film does exhibit a springlike resistance, which is dependent on the displacement of the shaft relative to the bushing, this force is not linearly related to the displacement, nor is it colinear with it. A fluid film also exhibits damping effects that play a very important role in the mechanical stability of this type of bearing.

In this section, we will find that fluid-film forces in hydrodynamic journal bearings tend to cause the bearing journal to orbit within the bearing bushing. We will show that the stability of the whirl orbit depends on the relative strength of the tangential fluid-film "stiffness" and the tangential fluid-film damping (Figure 5.26). From a simplistic fluid mechanic model of the lubricant film, we will attribute the negative tangential stiffness force and a tendency to whirl at one-half the shaft speed to a "surplus fluid effect" (Figure 5.27). Finally, we will show the results computed from a comprehensive fluid-film model (Figure 5.29), which maps the boundary between unstable (growing orbit radius) and stable (constant or decreasing orbit radius) bearing operation (Figure 5.30). Typical antiwhirl designs in common use are then illustrated (Figure 5.31).

We will begin our analysis of this very complex subject with simplified models of a rotor–bearing system. These models, one mechanical and one fluid, will contain the two important features of the system that are responsible for the unique behavior of this type of bearing. They are a fluid-film damping force and a tangential fluid-film force tending to drive the journal in the orbital motion within the bushing.

Figure 5.25 shows the journal with its center c eccentric relative to the center of the bushing, O. Fluid-film bearings require an eccentric journal to generate the nonsymmetric pressure distribution in the lubricant that carries the external load.

The distance between the bushing center and the shaft center is r. Under perfectly steady-load operating conditions, this eccentricity and its orientation remain constant, assuming a position such that the fluid-generated forces just balance the externally applied force F. However, in real applications, the applied force F usually fluctuates. The causes and form of these fluctuations vary from machine to machine, but we can safely assume that they will be there. Because of the special features of the fluid film mentioned before, it is possible that these fluctuations will set the journal into motion away from its steady-state operating position. If that happens, the

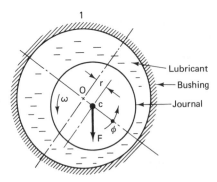

Figure 5.25 Cylindrical Journal Bearing in Bushing

journal may spiral out until it touches the bushing with disastrous results. On the other hand, it could orbit around the bushing center at a constant eccentricity, executing what is called a "limit cycle." The latter motion is certainly preferable. For that reason, we will focus our attention on determining the conditions that lead to that kind of motion.

Figure 5.26 shows the forces that are assumed to act on the journal. Forces F_r and F_ϕ are the components of the externally applied force F. Forces f_r and f_ϕ are the radial and tangential components, respectively, of the fluid-film forces assumed to act on the journal. The film has been idealized as springs and dashpots in the radial and tangential directions. The tangential "spring force" $k_{\phi r}r$ is shown acting in an unusual direction, i.e., assisting the tangential motion rather than opposing it. A "negative spring" of this type is contrary to our notions of an elastic stiffness, but can exist in fluid-film bearings with strange consequences, as we shall see.

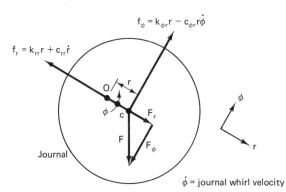

$\dot{\phi}$ = journal whirl velocity   **Figure 5.26** Forces on Journal

The kinetic energy of the shaft is given by

$$T = \tfrac{1}{2}m[\dot{r}^2 + (r\dot{\phi})^2] + \tfrac{1}{2}J\dot{\phi}^2$$

By neglecting the small potential-energy variables due to elevation changes, the Lagrangian is $L = T$. There are two degrees of freedom, r and ϕ, which yield two Lagrange equations:

$$\frac{d}{dt}\left(\frac{\partial L}{\partial \dot{r}}\right) - \frac{\partial L}{\partial r} = Q_r$$

and

$$\frac{d}{dt}\left(\frac{\partial L}{\partial \dot{\phi}}\right) - \frac{\partial L}{\partial \phi} = Q_\phi$$

Applying L to these equations yields

$$m\ddot{r} - mr(\dot{\phi})^2 = Q_r,$$

$$J\ddot{\phi} + mr^2\ddot{\phi} + 2r\dot{r}\dot{\phi}m = Q_\phi$$

On the right-hand side of these equations, the generalized force and torque are

$$Q_r = F_r - f_r = F_r - k_{rr}r - c_{rr}\dot{r}$$

$$Q_\phi = r(f_\phi - F_\phi) = -F_\phi r + k_{\phi r}r^2 - c_{\phi r}r^2\dot{\phi}$$

so that the dynamic equations become

$$m\ddot{r} - mr\dot{\phi}^2 + c_{rr}\dot{r} + k_{rr}r = F_r$$

$$J\ddot{\phi} + mr^2\ddot{\phi} + 2r\dot{r}\dot{\phi}m + c_{\phi r}r^2\dot{\phi} - k_{\phi r}r^2 = -F_\phi r$$

(5.44)

Our experience with the whirling elastic shaft suggests that a journal whirling at a constant radius might do so with a constant whirl velocity, i.e., $\ddot{\phi} = 0$. The same conclusion can be reached based on the symmetry of the fluid film, i.e., if the eccentricity remains constant, there are no variations in the film thickness that would cause $\dot{\phi}$ to vary.

If $\dot{\phi}$ is constant, the second homogeneous differential equation of motion then becomes

$$2\dot{r}\dot{\phi}m + c_{\phi r}r\dot{\phi} - k_{\phi r}r = 0$$

The solution of this equation is

$$r = r_o e^{at}$$

which requires that

$$a = \frac{k_{\phi r} - c_{\phi r}\dot{\phi}}{2m\dot{\phi}}$$

(5.45)

The radius of the journal orbit will grow, decay, or remain constant, depending on the magnitude of a. A constant-radius orbit, i.e., the boundary between stable and unstable operation, corresponds to $a = 0$ or $r = r_o$. If that solution is inserted into the first homogeneous differential equation, a formula for the constant-whirl angular velocity is obtained, which is

$$k_{rr} - m\dot{\phi}^2 = 0$$

or

$$\dot{\phi} = \sqrt{\frac{k_{rr}}{m}}, \quad a = 0$$

(5.46)

To summarize the results to this point, we have found that the journal modeled by the equations can whirl at a constant radius with a constant-whirl angular velocity $\dot{\phi}$ that is a natural frequency of the fluid-film–shaft–mass system. We have not yet ascertained the shaft speed at which this kind of whirl occurs. The empirical answer to this question was known to engineers long before a satisfactory physical explanation for it was obtained. To answer this question, we will again employ a very simplified model.

Figure 5.27 shows the lubricant flow around the clearance space of a journal bearing. The nominal clearance between the journal and its bushing, i.e., the difference in their radii, is c. When the journal is eccentric, the gap on one side of the

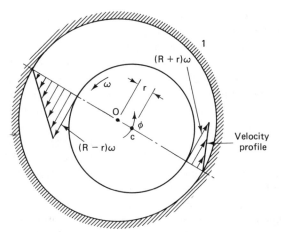

Figure 5.27 Lubricant Flow

journal is increased by r while the gap on the other side is decreased by r, as shown in the figure. When the eccentricity is small compared to the nominal clearance,[†] a reasonable approximation for the lubricant-velocity profile is a straight line. The journal's angular velocity is ω, so that the lubricant flow rate downward on the right side per unit-bearing length is

$$Q_R = \left(\frac{R - r}{2}\right)\omega(c + r)$$

while the flow rate upward on the left is

$$Q_L = \left(\frac{R + r}{2}\right)\omega(c - r)$$

Since the eccentricity is always very much smaller than the shaft radius, $R - r$ and $R + r$ are very nearly equal to R. Then the excess flow of lubricant into the lower half of the bearing in Figure 5.27 is

$$Q_R - Q_L = Rr\omega$$

Assuming that no lubricant leaks out the ends of the bearing, the journal must whirl upward to provide space for this excess fluid. The rate at which volume is made available by the orbiting journal is $2R\dot{\phi}r$. Equating the two flow rates to conserve the lubricant's mass yields

$$2R\dot{\phi}r = Rr\omega$$

or

$$\dot{\phi} = \omega/2 \qquad\qquad (5.47)$$

This equation says that the journal whirls at a constant radius with a frequency that is one-half the shaft speed, unlike the resonant whirl studied earlier, which occurs at

[†] Typical values of c/R range from 0.001 to 0.003.

the natural frequency of the elastic shaft–mass system. In practice, bearing whirl is actually observed at about 0.46 to 0.48 of shaft speed.

In addition to yielding a surprisingly accurate prediction of the journal whirl-speed to shaft-speed ratio, the bearing model of Figure 5.27 indicates the source of the sidewise destablizing force modeled as $k_{\phi r}$ in Figure 5.26. This force arises when leakage from the bearing ends is unable to accommodate the surplus pumped by the shaft. In the fluid model of Figure 5.27, the journal solved this problem by executing a steady, never-ending, self-excited circular whirl within the bushing.

At this point, one might well ask how does this behavior differ from the results obtained earlier with unbalanced shafts? Whirl due to unbalance, called "synchronous whirl" because it occurs at shaft speed, is present at all speeds and is most severe at the critical speeds of the mass–shaft system. "Fluid-film whirl," on the other hand, is not caused by unbalance, but by the fluid mechanics of the lubricant film. The whirl speed of the journal in its bushing is at a speed slightly less than one-half the shaft speed. If unstable, the whirl amplitude will grow rapidly, causing the journal and bushing to rub and destroy themselves.

Naturally, one would like to avoid bearing whirl if possible. The simplified mechanical model used before indicates that a stable solution exists if the proper relationship between the fluid-film stiffness, damping, and the shaft speed are achieved. To accurately locate the boundary between stable and unstable bearing operation, a more realistic model must be employed. To do this, we will consider the journal to be running in its static-equilibrium position, so that the fluid-film forces exactly cancel the steady external force applied to the shaft. We will then imagine that the journal is given an arbitrary small displacement from its equilibrium position, as shown in Figure 5.28. Displacements x and y are the horizontal and vertical components, respectively, of the shaft center. To accommodate the direct and cross-coupling fluid-film forces, the equations of Section 5.10 will be extended to read

$$m\ddot{x} + c_{xx}\dot{x} + c_{xy}\dot{y} + k_{xx}x + k_{xy}y = f_x$$
$$m\ddot{y} + c_{yy}\dot{y} + c_{yx}\dot{x} + k_{yy}y + k_{yx}x = f_y$$

(5.48)

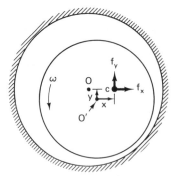

O = bushing center (fixed)
O' = shaft-equilibrium position
C = perturbed-shaft position

Figure 5.28 Perturbed Journal in Bushing

Figure 5.29 is a schematic drawing of the mechanical equivalent of the fluid film modeled by these equations. The forces f_x and f_y are the small disturbing-force components that give rise to x and y. The damping and stiffness coefficients, c and k, respectively, in these equations, are obtained by solving the complete bearing-fluid mechanic equations, called the Reynold's equations. Because of their complexity, these solutions are usually obtained with the aid of a computer.

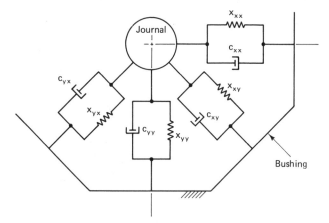

Figure 5.29   Mechanical Equivalent Fluid Film

Before attempting to solve these more comprehensive modeling equations, it is customary to introduce dimensionless coefficients, displacements and time. These are defined by

$$K_{xx} = \frac{k_{xx}}{F/c} \qquad C_{xx} = \frac{c_{xx}}{F/\omega c} \qquad \text{etc.}$$

and

$$X = \frac{x}{c} \qquad Y = \frac{y}{c} \qquad T = \omega t$$

where

  F = load applied to the journal
  ω = shaft speed
  c = bearing clearance

Applying these to the differential equations reduces them to

$$\frac{d^2 X}{dT^2} + S\left(C_{xx}\frac{dX}{dT} + C_{xy}\frac{dY}{dT} + K_{xx}X + K_{xy}Y\right) = \frac{f_x}{cm\omega^2}$$

$$\frac{d^2 Y}{dT^2} + S\left(C_{yy}\frac{dY}{dT} + C_{yx}\frac{dX}{dT} + K_{yy}Y + k_{yx}X\right) = \frac{f_y}{cm\omega^2}$$

where

$$S = F/mc\omega^2$$

Except for S, all the coefficients in this equation are fixed once the bearing's equilibrium position within the bearing is established. In hydrodynamic bearing theory, the radial displacement of the journal center is called its eccentricity. In the simplified model, this radius was denoted by r, since e had been previously used to indicate rotor unbalance. We will continue to use r as the journal's eccentricity, although e is commonly used in bearing technology for this quantity.

The characteristic equation for this system can be formed by assuming solutions to the differential equations in the form

$$X = A_1 e^{\omega t} \qquad Y = A_2 e^{\omega t}$$

so that the homogeneous differential equations reduce to

$$[\omega^2 + S(K_{xx} + C_{xx}\omega)]A_1 + [S(K_{xy} + C_{xy}\omega)]A_2 = 0$$
$$[S(K_{yx} + (C_{yx})\omega]A_1 + [\omega^2 + S(K_{yy} + C_{yy}\omega)]A_2 = 0 \qquad (5.49)$$

The characteristic equation that results from setting the determinant of the coefficients equal to zero is a fourth-order polynomial in ω of the form

$$\omega^4 + B_3\omega^3 + B_2\omega^2 + B_1\omega + B_o = 0$$

Unlike the previous cases where damping was not included, the roots of this polynomial are complex instead of purely imaginary, i.e.,

$$\omega_{1,2} = \alpha_1 \pm i\beta_1$$
$$\omega_{3,4} = \alpha_2 \pm i\beta_2$$

If the real part of either of the complex conjugates, α_1 and α_2, is positive, the transient displacements grow without bound. We interpret that to mean that given a small perturbation from its equilibrium position, the journal tends to continue moving rather than returning to its equilibrium position. The journal may move to a larger eccentricity and then orbit at a constant-radius limit cycle, such as we found from the simplified model. On the other hand, the radius may continue to grow until the journal touches the bushing. To be safe, we would rather not risk either of these outcomes, in which case S must be adjusted to that $\alpha_1 < 0$ and $\alpha_2 < 0$. We can determine the boundary between the stable and unstable motion by applying Routh's stability criterion. Figure 5.30 shows the result. To test a bearing's stability, one calculates the eccentricity ratio, r/c, from bearing theory. Then, with the stability parameter,[†] a coordinate point in the map is found and stability or lack of it is established.

For eccentricity ratios, r/c, larger than 0.8, Figure 5.30 shows that a cylindrical journal bearing is stable under all operating conditions. Because the shaft touches

---

[†] When the load F is the dead weight of the rotor, then $S = g/c\omega^2$

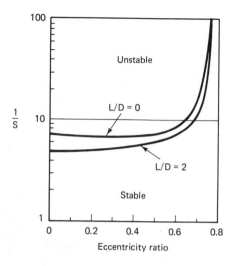

Figure 5.30 Typical Cylindrical Journal Bearing Stability Map (L = Length, D = Diameter)

the bushing when $r/c = 1.0$, this leaves a very small range of eccentricities where stability is absolutely guaranteed.

Because of this deficiency, a number of antiwhirl bearing designs have been developed to replace the whirl-prone cylindrical journal bearing. Some of these are illustrated in Figure 5.31. Each of these configurations lacks the symmetry of clearance space that, in the plain cylindrical bearing (a), permits whirl to continue. Recall that in the model of Figure 5.27, the journal can never stop whirling, because as the shaft moves on its orbit, it always sees the same clearance-space geometry. As the shaft moves in the antiwhirl bearings, it encounters new geometric configurations for its clearance space. Because of this, the journal may lock in to a position where the tendency to whirl disappears.

Of these designs, the elliptical is the most popular because it can be made from the two halves of a cylindrical journal bearing. The most expensive is the tilting-pad bearing. It also has the interesting property that k_{xy} and k_{yx} are zero. None of these bearings is absolutely stable. They just have larger stable regions on their stability maps than the plain cylindrical bearing shown in Figure 5.30.

We can summarize our findings in this section as follows:

1. Cylindrical fluid-film bearings have a tendency to whirl, i.e., the shaft can orbit within the bearing bushing.

2. A simplistic analysis indicates that this tendency is caused by a lubricant-fluid negative tangential spring effect, $k_{\phi r}$ of Figure 5.26.

3. A simplistic model (Figure 5.27) shows that the tangential force propelling the shaft around its orbit can be attributed to a surplus of lubricant. Mass conservation requires that this whirl speed, $\dot{\phi}$, be at one-half the shaft speed, ω.

4. According to the model of Figure 5.27, cylindrical journal bearings always whirl. That is not true. It is true that when they do whirl, they do so with $\dot{\phi}$ slightly less than $\frac{1}{2}\omega$.

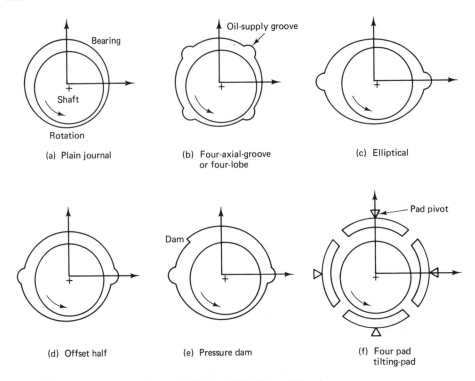

Figure 5.31   Anti-Whirl Journal Bearings

5. The more elaborate model of Figure 5.29 shows that nonwhirling (stable) operation *can* be obtained if the bearing number ($S = F/mc\omega^2$) is large or the shaft eccentricity is large (Figure 5.30).

6. The alternative to the cylindrical bearing is an antiwhirl design; some are shown in Figure 5.31.

Before leaving this section, it would probably be helpful if its results were compared with those of Sections 5.7 and 5.8. In those sections, we were concerned with unbalanced elastic shafts. Whirl of the bowed shaft at shaft speed, ω, called "synchronous whirl" occurs at all speeds but is most severe when the shaft speed is near or at one of the shaft/disk natural frequencies. The bearings play no role in this phenomenon.

The fluid-film whirl discussed in this section also occurs at all speeds, but it is not at the shaft speed. This bearing-caused whirl is at a speed slightly less than one-half shaft speed. When this whirl is unstable, the shaft eccentricity grows without bound, so that the shaft touches the bushing, causing a loss of lubricant film and the ultimate destruction of the bearing.

Obviously, rolling-element bearings are not subject to fluid-film whirl. They do experience resonant whirl, however. The elasticity of the rolling elements has a softening effect on the equivalent stiffness of the shaft–bearing system.

5.12 MECHANICS OF RIGID-ROTOR BALANCING

Figure 5.32 shows a rigid rotor whose center of mass g is not at its geometric center c. Moving coordinates $x'y'z'$ are fixed to the rotor at g. The primes are included to indicate that the origin of these coordinates is at the center of mass. The unbalance distance e is measured normal to the shaft axis z. The angular velocity and angular acceleration of the shaft are ω_z and $\dot{\omega}_z$, respectively. The moment-of-momentum equations from Section 3.3 are

$$M_{x'} = I_{x'z'}\dot{\omega}_z + I_{y'z'}\omega_z^2$$

$$M_{y'} = -I_{y'z'}\dot{\omega}_z - I_{z'x'}\omega_z^2 \qquad (5.50)$$

$$M_{z'} = I_{z'z'}\dot{\omega}_z$$

where the external moments, $M_{x'}$, $M_{y'}$, and $M_{z'}$ are taken about the rotating $x'y'z'$ axes. These moments are applied to the rotor ends by the bearings in the form of rotating forces or couples.

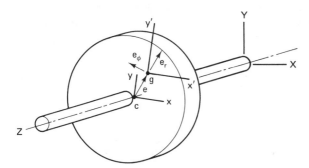

Figure 5.32 Unbalanced Rigid Rotor

Newton's second law for the rotor is

$$\Sigma\mathbf{F} = m(e\dot{\omega}_z\mathbf{e}_\phi - e\omega_z^2\mathbf{e}_r) \qquad (5.51)$$

where ΣF is the vector sum of the forces applied by the bearings.

Except for the dead-weight contributions, these forces and moments vanish when the shaft stops rotating. The sum of the rotating forces exerted on the bearings as reactions, i.e. ΣF, vanish when the balance distance e is zero. However, as long as the products of inertia, $I_{x'z'} = I_{z'x'}$ and $I_{y'z'} = I_{z'y'}$, are nonzero, the bearings will experience rotating reactive moments, $M_{x'}$ and $M_{y'}$, which will cause the machine to vibrate. The torque M_z is required to accelerate or decelerate the rotor, but contributes nothing to the vibration of the machine.

To achieve a perfect dynamic balance requires that we simultaneously reduce the products of inertia and the unbalance to zero. The unbalance distance e can be reduced to zero by positioning a single balance mass anywhere along the rotor, so that the resulting center of mass is on the shaft axis. The shaft would then be "statically" balanced. Unfortunately, the products of inertia cannot be reduced to zero by the addition of a single balance weight. Two balance masses must be added in two different balance planes, as shown in Figure 5.33.

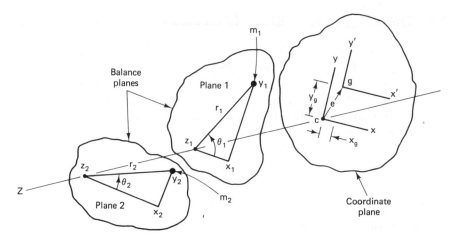

Figure 5.33   Balance Planes with Balance Weights

The coordinates of these planes relative to c along the axis of the shaft are z_1 and z_2. The coordinates of the balance masses in these planes are x_1, y_1, and x_2, y_2.

After the small balance masses m_1 and m_2 are applied, the x coordinate of the new center of mass is calculated from the ratio

$$\frac{m_1 x_1 + m_2 x_2 + m x_g}{m_1 + m_2 + m} \qquad m_1, m_2 \ll m$$

but since the center of mass is to lie on the shaft axis, this ratio will be set equal to zero. Hence,

$$m_1 x_1 + m_2 x_2 + m x_g = 0$$

The y coordinate of the new center of mass will also be equal to zero if

$$m_1 y_1 + m_2 y_2 + m y_g = 0$$

The static-balance conditions are satisfied by these equations. The center of mass, originally at g, is now at c on the Z axis.

If the center of mass is now at c in Figure 5.33, the axial coordinate z of each particle is unchanged. According to the parallel-axis theorem, the products of inertia of the rotor *without* the balance masses about the new mass-center coordinates xyz are the same as for the old coordinates $x'y'z'$, i.e.,

$$I_{x'z'} = I_{xz} \qquad I_{y'z'} = I_{yz}$$

The products of inertia about the new mass-center coordinates *with* the balance masses attached are

$$(I_{xz})_{\text{total}} = I_{x'z'} + m_1 x_1 z_1 + m_2 x_2 z_2$$
$$\qquad\qquad\qquad\qquad\qquad m_1, m_2 \ll m \qquad\qquad (5.52)$$
$$(I_{yz})_{\text{total}} = I_{y'z'} + m_1 y_1 z_1 + m_2 y_2 z_2$$

Dynamic balance requires that $(I_{xz})_{\text{total}} = (I_{yz})_{\text{total}} = 0$.

The set of equations that yields perfect dynamic balance are, therefore:

$$(m_1 x_1) + (m_2 x_2) + m x_g = 0$$

$$(m_1 y_1) + (m_2 y_2) + m y_g = 0 \tag{5.53a}$$

$$(m_1 x_1)z_1 + (m_2 x_2)z_2 + I_{x'z'} = 0$$

$$(m_1 y_1)z_1 + (m_2 y_2)z_2 + I_{y'z'} = 0 \tag{5.53b}$$

For practical reasons, the balance planes cannot be chosen arbitrarily. Balance masses are usually added to or subtracted from the rotor at locations where they will not interfere with its operation. When the position of the balance planes, z_1 and z_2, are specified, the four previous equations become simultaneous equations for the four quantities $m_1 x_1$, $m_2 x_2$, $m_1 y_1$, and $m_2 y_2$. The radial location of the balance masses in plane 1 is given by

$$(m_1 x_1)^2 + (m_1 y_1)^2 = m_1^2 r_1^2 \tag{5.54a}$$

Since the left side of this equation is known, mass m_1 can be found by choosing the radial location r_1. Balance mass m_2 is likewise determined by selecting r_2 in balance plane 2, so that

$$m_2^2 = \frac{1}{r_2^2}[(m_2 x_2)^2 + (m_2 y_2)^2] \tag{5.54b}$$

The angular location relative to the x axis of each balance mass is given by

$$\tan \theta_1 = \frac{m_1 y_1}{m_1 x_1}$$

$$\tag{5.55}$$

$$\tan \theta_2 = \frac{m_2 y_2}{m_2 x_2}$$

The results of our investigation of rigid-rotor balancing can be summarized as follows:

1. Two balance masses applied in two arbitrarily chosen balance planes are required.
2. The balance masses can be applied at any nonzero radii.
3. The sizes of the balance masses are determined once the balance planes and radii are selected.
4. The angular location of the balance masses is determined by the choice of the balance planes.

As an example of how to apply Equations (5.53), consider the unbalanced rotor shown in Figure 5.34. The unbalance of this rotor is caused by two small weights of 0.1 lb$_f$ attached to the rotor, as shown in the figure. From symmetry, we know that

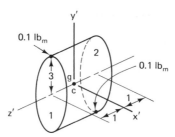

Figure 5.34   Unbalanced Rotor

g lies on the z axis midway between the ends of the rotor, so that $x_g = y_g = 0$. One simple way to balance this rotor may be obvious based on symmetry. The solution of Equations (5.53) will substantiate that observation.

Since both unbalances lie in the yz plane, $I_{x'z'} = 0$, and

$$I_{y'z'} = \frac{0.1}{g}(1)(3) + \frac{0.1}{g}(-1)(-3) = \frac{0.6}{g}$$

Balance planes 1 and 2 are marked in Figure 5.34, yielding $z_1 = 1.0$ and $z_2 = -1.0$. Substituting this data into Equations (5.53) yields

$$m_1 x_1 + m_2 x_2 + 0 = 0$$

$$m_1 y_1 + m_2 y_2 + 0 = 0$$

$$m_1 x_1 - m_2 x_2 + 0 = 0$$

$$m_1 y_1 - m_2 y_2 + \frac{0.6}{g} = 0$$

The first and third equations can only be satisfied if $m_1 x_1 = m_2 x_2 = 0$. Adding the second and fourth equations yields

$$m_1 y_1 = -\frac{1}{2}\frac{0.6}{g}$$

and subtracting the second and fourth equations gives

$$m_2 y_2 = \frac{1}{2}\frac{0.6}{g}$$

Since m_1 and m_2 are not necessarily zero, then $x_1 = x_2 = 0$. If we choose $m_1 = m_2 = 0.1$, then

$$y_1 = -3.0$$

$$y_2 = 3.0$$

This is the solution we might have proposed based on symmetry considerations, but our analysis shows that this is not the only solution.

5.13 BALANCING TECHNIQUE: SHORT ROTOR

In practice, one rarely knows x_g, y_g, $I_{x'z'}$, and $I_{y'z'}$ for a rotor, so that the theoretical equations can seldom be used. On the other hand, analysis lays the foundations for the balancing techniques to be described in this section.

Figure 5.35(a) shows a short rotor unbalanced by the exciting force $me\omega^2$ obtained from the analysis of Section 5.10. The force $me\omega^2$ is drawn on the face of the disk passing through point a. A short rotor has negligible products of inertia. Equations (5.53b) are satisfied in this case by letting $z_1 = z_2 = 0$, which means that only a single balance weight in the plane of the disk is needed. The simple technique used to balance this rotor will subsequently be extended to a two-plane method for long disks with nonnegligible products of inertia.

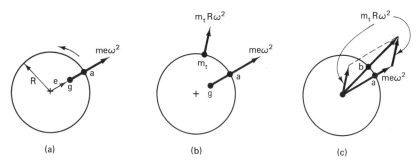

(a)   (b)   (c)

Figure 5.35   Effect of Trial Mass

Figure 5.35(b) shows the additional centrifugal force component that results from attaching a trial mass, m_t, to the rim of the disk. The resultant of these forces is drawn on the face of the disk passing through b in Figure 5.35(c).

Figure 5.36 shows a device used to locate points a and b. If the unbalanced rotor is supported in the balancing device shown in the figure, the sinusoidal variation in the gap, δ_1, between the disk and a fixed reference can be measured. The size of this variation at a given speed, δa, is recorded and the location of the minimum gap, a, is marked on the disk. When a trial mass m_t is added to the rim, the new location

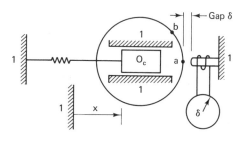

Figure 5.36 Experimental Balancing Device

of the minimum gap, *b*, is also noted along with the size of that variation, **δb**. These experimental results are combined vectorially, as shown in Figure 5.37, where **δt** is the vector corresponding to the centrifugal contribution of the balance mass m_t, i.e., $m_t R\omega^2$ of Figure 5.35(c).

If the trial-mass location is now moved through angle ϕ, the two centrifugal vectors will be oppositely directed. The balance mass m_b that causes these vectors to cancel is obtained from the scaling equation:

$$m_b = \left|\frac{\delta a}{\delta t}\right| m_t$$

Figure 5.38 shows the final configuration.

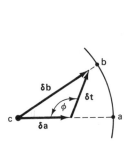

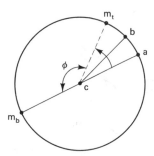

Figure 5.37 Effect of   **Figure 5.38**   Balanced Disk
Trial Mass—Measured

The experimental procedure described can be written mathematically rather neatly using complex numbers. The vectors needed are shown in the complex plane of Figure 5.39.

Each of these vectors is written in complex form as follows:

$$\delta a = \delta a e^{i\theta_a} \qquad \mathbf{m}_t = m_t e^{i\theta_t}$$

$$\delta b = \delta b e^{i\theta_b} \qquad \delta t = \delta b - \delta a$$

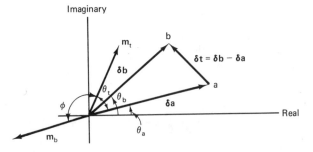

Figure 5.39 Balancing Vectors in the Complex Plane

As we have seen, balance is obtained by rotating and scaling δt so that it cancels δa. This transformation will be accomplished by the vector $\mathbf{T}$, i.e.,

$$-\delta a = \mathbf{T} \, \delta t$$

where

$$\mathbf{T} = \frac{-\delta \mathbf{a}}{\delta \mathbf{b} - \delta \mathbf{a}}$$

$$= \frac{1}{1 - \delta \mathbf{b}/\delta \mathbf{a}} \tag{5.56}$$

$$= \frac{1}{1 - (\delta b/\delta a)e^{i(\theta_b - \theta_a)}}$$

The rotation and scaling of the trial-mass vector is accomplished using this vector, i.e.,

$$\mathbf{m}_b = m_b e^{i(\theta_t + \phi)} = \mathbf{T}\mathbf{m}_t$$

$$= \frac{m_t e^{i\theta_t}}{1 - (\delta b/\delta a)e^{i(\theta_b - \theta_a)}} \tag{5.57}$$

The right side of this equation contains the measured quantities. Its modulus is the required balance mass and its argument contains the phase angle ϕ.

To illustrate this technique, consider a 2.0-in. diameter thin disk mounted in the balancing machine of Figure 5.36. Suppose that, unknown to us, this disk is unbalanced by a small 0.01 lb$_m$ mass on the rim. When the disk is at rest ($\omega = 0$), we set the gap δ to 0.020 in. Following that, we run the rotor up to a set speed and note the smallest gap is 0.010 in. At the point of closest approach, we mark the disk with an a. Starting at point a, we mark off the disk face in degrees. Now we have $\delta a = 0.01$ in. at $\theta_a = 0$ (Figure 5.39).

Assume that we arbitrarily attach a 0.005-lb$_m$ trial mass (m_t) at 1.0 in. and 45° and run the rotor up to the *same* set speed again. The narrowest gap will now be 0.006 in. at $\theta_b = 14.6°$. We calculate $\delta b = 0.014$ in. at point b in Figure 5.35(c) and insert all the measured data in Equation (5.57), so that

$$\mathbf{m}_b = \frac{0.005 e^{i45°}}{1 - (0.014/0.01)e^{i14.6°}}$$

$$= 0.01 e^{-180i}$$

We interpret this result to mean that balance will be obtained if we place a 0.01-lb$_m$ mass directly opposite to a, i.e., 180° away.

The same result would have been obtained if the unknown unbalance had been the result of 0.02-lb$_m$ mass at a 0.5-in. radius, since the product me is the same for both cases. The balance achieved in this case is actually equivalent to a "static" balance since the products of inertia play no role.

5.14 BALANCING TECHNIQUE: LONG ROTOR

The complex-variable method can be extended to the two-plane balancing technique required to balance a long rotor, such as shown in Figure 5.40. The balance planes are labeled 1 and 2 in this figure. The rotor is first run at some ω, preferably close to its operating speed, and δa_1 and δa_2 determined in each balance plane. A trial mass m_{t1} is then attached in balance plane 1 at θ_{t1} and δb's measured in both balance planes, since m_{t1} affects the unbalance throughout the rotor. These vectors will be noted as δb_{11} in plane 1 and δb_{21} in plane 2. Figure 5.41 shows the vector diagrams in both planes.

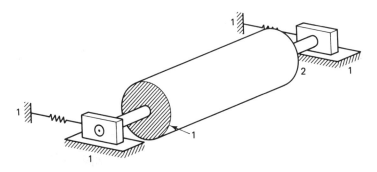

Figure 5.40   Unbalanced Long Rotor

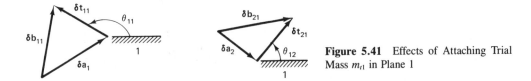

Figure 5.41   Effects of Attaching Trial Mass m_{t1} in Plane 1

Vector δt_{21} is the effect trial mass m_{t1} has in balance plane 2. Since δt_{11} and δt_{21} result from the same cause, m_{t1}, they are also related to each other. Vector $\mathbf{I}_1$ will be defined as the vector that transforms δt_{11} into δt_{21}, i.e.,

$$\mathbf{I}_1\, \delta t_{11} = \delta t_{21}$$

or

$$\mathbf{I}_1 = \frac{\delta t_{21}}{\delta t_{11}} e^{(\theta_{21} - \theta_{11})i} \tag{5.58a}$$

The vector $\mathbf{I}_1$ is a property of the rotor that is obtained experimentally.

By repeating the same procedure, applying a trial mass m_{t2} in balance plane 2 at θ_{t2} results in δt_{12} and δt_{22} in planes 1 and 2, respectively. The vector relating these two vectors is defined by

$$\mathbf{I}_2 \, \delta t_{22} = \delta t_{12}$$

or

$$\mathbf{I}_2 = \frac{\delta t_{12}}{\delta t_{22}} e^{(\theta_{12} - \theta_{22})i} \tag{5.58b}$$

The transformation vectors that scale and rotate the trial masses into the balance masses will be designated as $\mathbf{T}_1$ and $\mathbf{T}_2$. In plane 1, the result of rotating and scaling m_{t1} is $\mathbf{T}_1 \, \delta t_{11}$. The result of rotating and scaling m_{t2} in plane 2 is $\mathbf{T}_2 \, \delta t_{22}$.

These are the direct effects in the planes where the changes are made. However, vector $\mathbf{T}_1 \, \delta t_{11}$ in plane 1 has a simultaneous effect in plane 2 given by $\mathbf{I}_1 \mathbf{T}_1 \, \delta t_{11}$, and, conversely, vector $\mathbf{T}_2 \, \delta t_{22}$ in plane 2 has a simultaneous effect in plane 1 of $\mathbf{I}_2 \mathbf{T}_2 \, \delta t_{22}$.

The total effect of rotating and scaling both trial masses are, in plane 1,

$$\mathbf{T}_1 \, \delta t_{11} + \mathbf{I}_2 \mathbf{T}_2 \, \delta t_{22}$$

and, in plane 2,

$$\mathbf{T}_2 \, \delta t_{22} + \mathbf{I}_1 \mathbf{T}_1 \, \delta t_{11}$$

Complete balance is attained when these vectors cancel the original unbalances, i.e.,

$$\mathbf{T}_1 \, \delta t_{11} + \mathbf{T}_2 \mathbf{I}_2 \, \delta t_{22} + \delta \mathbf{a}_1 = 0$$

$$\mathbf{T}_1 \mathbf{I}_1 \, \delta t_{11} + \mathbf{T}_2 \, \delta t_{22} + \delta \mathbf{a}_2 = 0$$

These are two simultaneous equations for vectors $\mathbf{T}_1$ and $\mathbf{T}_2$ that transform the trial masses into the balance masses. Solving these two equations yields

$$\mathbf{T}_1 = \frac{\delta t_{12} \, \delta \mathbf{a}_2 - \delta t_{22} \, \delta \mathbf{a}_1}{\delta t_{11} \, \delta t_{22} - \delta t_{12} \, \delta t_{21}}$$

$$\mathbf{T}_2 = \frac{\delta t_{21} \, \delta \mathbf{a}_1 - \delta t_{11} \, \delta \mathbf{a}_2}{\delta t_{11} \, \delta t_{22} - \delta t_{12} \, \delta t_{21}}$$

The right sides of these equations consist entirely of measured vectors. The balance-mass vectors $\mathbf{m}_{b1}$ and $\mathbf{m}_{b2}$ are obtained from the trial-mass vectors:

$$\mathbf{m}_{t1} = m_{t1} e^{i\theta_{t1}}$$

$$\mathbf{m}_{t2} = m_{t2} e^{i\theta_{t2}}$$

By using $\mathbf{T}_1$ and $\mathbf{T}_2$,

$$\mathbf{m}_{b1} = \mathbf{T}_1 m_{t1} e^{i\theta_{t1}}$$

$$\mathbf{m}_{b2} = \mathbf{T}_2 m_{t2} e^{i\theta_{t2}} \tag{5.59}$$

As an example, consider the following hypothetical case. Preliminary tests on a balance machine show minimum gaps at 69.4° in plane 1 and 161.6° in plane 2.

The maximum horizontal displacements measured relative to the rest positions are 0.0085 and 0.0032 inches, respectively. The displacement vectors are, therefore,

$$\delta\mathbf{a}_1 = 0.0085e^{69.4i}$$

$$\delta\mathbf{a}_2 = 0.0032e^{161.6i}$$

One-tenth-pound trial-balance masses are attached successively to planes 1 and 2 at 90° relative to the reference line. They are expressed in complex form as

$$\mathbf{m}_{t1} = \mathbf{m}_{t2} = 0.1e^{90i}$$

Assume that the vectors calculated from tests with the trial masses are

$$\delta\mathbf{t}_{11} = 0.0054e^{248.2i}$$

$$\delta\mathbf{t}_{21} = 0.0016e^{55.3i}$$

$$\delta\mathbf{t}_{22} = 0.0022e^{-74.1i}$$

$$\delta\mathbf{t}_{12} = 0.0016e^{55.3i}$$

(These results were chosen arbitrarily for purposes of illustrating the technique. Actually, they are not independent, but are related to each other through the state of unbalance.)

Since the equations for $\mathbf{T}_1$ and $\mathbf{T}_2$ are ratios, we can simplify calculations by writing

$$\mathbf{T}_1 = \frac{(1.6e^{55.3i}3.2e^{161.6i}) - (2.2e^{-74.1i}8.5e^{69.4i})}{(5.4e^{248.2i}2.2e^{-74.1i}) - (1.6e^{55.3i}1.6e^{55.3i})}$$

$$= \frac{5.1e^{216.9i} - 18.7e^{-4.7i}}{11.9e^{174.1i} - 2.6e^{110.6i}}$$

$$= \frac{-2.3i - 1.5i}{-10.9 - 1.2i} = \frac{22.7e^{183.9i}}{11.0e^{186.2i}}$$

$$= 2.1e^{-2.3i}$$

The balance-mass vector is

$$\mathbf{m}_{b1} = \mathbf{T}_1\mathbf{m}_t = 2.1e^{-2.3i}0.1e^{90i}$$

$$= 0.21e^{87.7i}$$

This shows that a 0.21-lb balance mass should be placed at 86.7° in balance plane 1.
For balance plane 2,

$$\mathbf{T}_2 = \frac{(1.6e^{55.3i}8.5e^{69.4i}) - (3.2e^{161.6i}5.4e^{248.2i})}{11.0e^{186.2i}}$$

$$= \frac{13.6e^{124.7i} - 17.3e^{49.8i}}{11.0e^{186.2i}}$$

$$= 1.7e^{-0.1i}$$

and

$$\mathbf{m}_{b2} = 1.7e^{-0.1i}0.1e^{90i}$$

$$= 0.17e^{89.9i}$$

so that balance is obtained also by adding a 0.17-lb balance mass at 89.9° in balance plane 2.

Nowadays, these complex-algebra computations are performed by a small computer attached to the balancing machine. The test data obtained from the trial runs are automatically reduced and the size and location of balance weights outputted.

The previous procedure is applicable when the rotor is run at less than about one-half of its first critical speed. At speeds higher than that, the rotor will bend, setting up centrifugal forces in addition to those that caused the original unbalance. Unlike "rigid" rotors, which can be balanced for all operating speeds, flexible rotors can only be balanced in two planes for one specific speed. They will be unbalanced for all others. A complete description of flexible-rotor balancing techniques is beyond the scope of this text.

Perfect balance is rarely, if ever, achieved. The cost of balancing increases with the quality of the balance. Some typical industry standards for balance grades are given in Figures 5.42 and 5.43 (see pages 218–21). The various grades are determined by the product of the unbalanced distance e and shaft speed ω. For example, suppose a shaft is rotating at 100 radians per second with a 2.5-mm unbalance. Then $e\omega = 100$, and the grade is G250. This unbalance is also about 0.1 inches and the shaft speed about $n = 1000$ rpm. Using these as coordinates in Figure 5.42, we also find that the balance grade is G250.

REFERENCES

1. Shigley, J. E., and Mitchell, L. P. *Mechanical Engineering Design,* 4th Ed., Chap. 12. New York: McGraw-Hill, 1983.
2. Excerpted from ANSI S2.19–1975 (ASA Standard 2–1975). Balance Quality of Rotating Rigid Bodies, Standards Secretarial, Acoustical Society of America, 335 East 45th Street, New York, New York 10017.

Balance quality grades G	$e\omega$ [a,b] (mm/sec)	Rotor types—General examples
G 4 000	4 000	Crankshaft-drives[c] of rigidly mounted slow marine diesel engines with uneven number of cylinders.[d]
G 1 600	1 600	Crankshaft-drives of rigidly mounted large two-cycle engines.
G 630	630	Crankshaft-drives of rigidly mounted large four-cycle engines. Crankshaft-drives of elastically mounted marine diesel engines.
G 250	250	Crankshaft-drives of rigidly mounted fast four-cylinder diesel engines.[d]
G 100	100	Crankshaft-drives of fast diesel engines with six or more cylinders.[d] Complete engines (gasoline or diesel) for cars, trucks, and loco-motives.[e]
G 40	40	Cars wheels, wheel rims, wheel sets, drive shafts Crankshaft-drives of elastically mounted fast four-cycle engines (gasoline or diesel) with six or more cylinders.[d] Crankshaft-drives for engines of cars, trucks, and locomotives.
G 16	16	Drive shafts (propeller shafts, cardan shafts) with special requirements. Parts of crushing machinery. Parts of agricultural machinery. Individual components of engines (gasoline or diesel) for cars, trucks, and locomotives. Crankshaft-drives of engines with six or more cylinders under special requirements. Slurry or dredge pump impeller.
G 6.3	6.3	Parts or process plant machines. Marine main turbine gears (merchant service). Centrifuge drums. Fans. Assembled aircraft gas turbine rotors. Fly wheels. Pump impellers. Machine-tool and general machinery parts. Normal electrical armatures. Individual components of engines under special requirements.

Balance quality grades G	$e\omega$ [a,b] (mm/sec)	Rotor types—General examples
G 2.5	2.5	Gas and steam turbines, including marine main turbines (merchant service). Rigid turbo-generator rotors. Rotors. Turbo-compressors. Machine-tool drives. Medium and large electrical armatures with special requirements. Small electrical armatures. Turbine-driven pumps.
G 1	1	Tape recorder and phonograph (gramophone) drives. Grinding-machine drives. Small electrical armatures with special requirements.
G 0.4	0.4	Spindles, disks, and armatures of precision grinders. Gyroscopes.

[a] $\omega = 2\pi n/60 \approx n/10$, if n is measured in revolutions per minute and ω in radians per second.

[b] In general, for rigid rotors with two correction planes, one-half of the recommended residual unbalance is to be taken for each plane: these values apply usually for any two arbitrarily chosen planes, but the state of unbalance may be improved upon at the bearings. (See Secs. 3.2 and 3.4.) For disk-shaped rotors the full recommended value holds for one plane. (See Sec. 3.)

[c] A crankshaft-drive is an assembly which includes the crankshaft, a flywheel clutch, pulley, vibration damper, rotating portion of connecting rod, etc. (See Sec. 3.5.)

[d] For purposes of this Standard, slow diesel engines are those with a piston velocity of less than 9 m/sec; fast diesel engines are those with a piston velocity of greater than 9 m/sec.

[e] In complete engines, the rotor mass comprises the sum of all masses belonging to the crankshaft-drive described in Note c above.

Figure 5.42

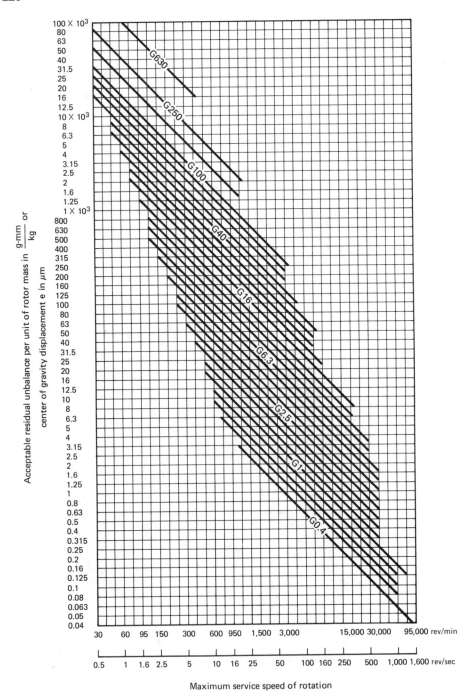

Figure 5.43

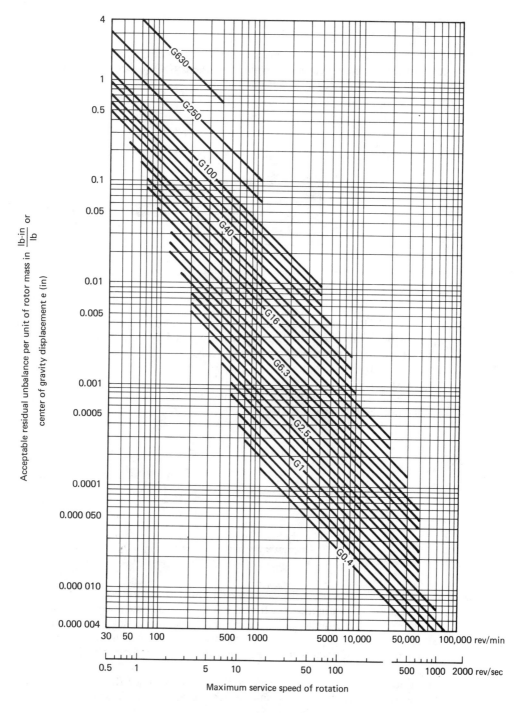

Figure 5.43 (Cont.)

PROBLEMS

5.1. For the system shown in the diagram, determine the following:
   (a) Using the Lagrange equations, obtain the differential equations of motion.
   (b) Find the natural frequencies for this system.

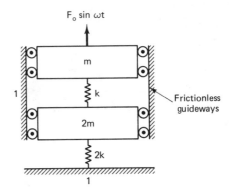

5.2. For the elastically supported machine shown in the diagrams, derive the differential equations of motion using the Lagrange equations and find the system's natural frequencies when g is constrained to move vertically.
   $k_1 = 2400$ lb$_f$/ft $l_1 = 4.5$ ft
   $k_2 = 2600$ lb$_f$/ft $l_2 = 5.5$ ft
   $w = 3220$ lb$_f$ radius of gyration from $g = 4$ ft
   Assume that the tilt angle is $< 10°$.

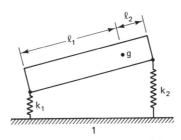

5.3. The diagram shows a rotor disk with an elastically constrained mass m_2 that can only move radially in a slot in the disk. It models a rotor–shaft system with a movable part. Length $r_o =$ spring unstretched length. For this model (neglecting gravity effects),
   (a) Obtain the equations of motion for the system when $\omega =$ constant.
   (b) From the equations of motion, determine the conditions for stable operation of the system.
   (c) Identify the "force" that gives rise to the disturbing torque T_s.
   (d) Determine the frequencies of the disturbing torque T_s during steady-state ($\omega =$ constant) stable operation.

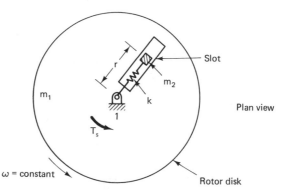

5.4. The machine shown in the diagram is supported on *four* k_1 springs and constrained by four k_2 springs. The rotor (2) rotates at $\omega_2 = 60$ Hz. Assuming that the frame and rotor are rigid and rigidly connected, determine the natural frequencies of the system. Comment on your results.

$k_1 = 5(10^5)$ lb$_f$/in.

$k_2 = 5(10^5)$ lb$_f$/in.

$W_1 = 600$ lb$_f$

$W_2 = 400$ lb$_f$

$J_{g1} = 150$ lb$_f$-in.-s^2 (frame)

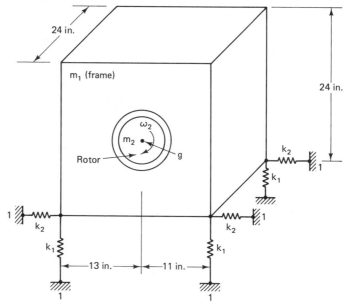

5.5. The diagram shows a rotor m_2 that is rigidly attached to mass m_1 at g_1, which is the center of mass of m_1. The rotor's center of mass is at g_2. The rotor rotates with a constant angular velocity of ω. Mass m_1 slides without friction in the horizontal guideways, restrained by spring k.

(a) Obtain the algebraic equation for the steady-state displacement of mass m.

(b) Determine the shaft speed ω at which resonance would occur.

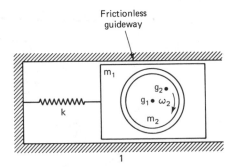

5.6. The mass m_1 shown in the diagram contains an unbalanced rotor m_2 that rotates at a constant speed of $\omega = \dot{\theta}$. It causes m_1 to oscillate up and down inside the vertical frictionless guideways. Assume that the vertical displacement of m_1 is measured from rest and that the potential-energy variations caused by the unbalance distance e are small.

   (a) Obtain the equation for T of the system.
   (b) Obtain the equation for V of the system
   (c) Using the appropriate Lagrange equation, obtain the differential equation for the vertical motion of the system.
   (d) Solve for the steady amplitude of the vertical response.

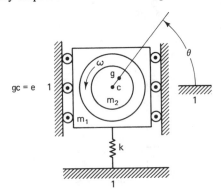

5.7. The tub of a washer is constrained by three equal springs k of equal lengths l, as shown in the diagram. They are 120° apart and connect the tub to ground. When the tub is centered, the springs are unloaded. During the spin cycle, when water is centrifuged from the wash, the drum rotates within the tub at a constant speed ω. The center of the wash–drum mass becomes eccentric a distance e away from c. This rotating unbalance causes the wash–drum–tub assembly to vibrate, displacing c from its rest position with a radial distance $r(t)$ and angle $\theta(t)$.

   (a) Show that when $r/l \ll 1.0$, the restoring force of the springs is given by

$$F_s = 3/2\ kr$$

   (b) Using the Lagrange equations and xy generalized coordinates, obtain the equations of motion.
   (c) Assuming that the tub has negligible mass, the wash and drum weigh 50 lb$_f$, $\omega = 400$ rpm, and the unbalance is 20 lb-in., calculate the value of k so that ω is three times the highest natural frequency.
   (d) Determine r for this operating condition.

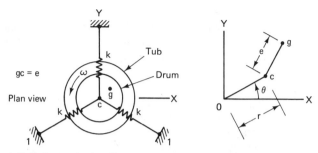

5.8. The shaft and disk shown in the diagram turn at a constant speed ω. The torque T_{bearings} required is that needed to overcome the constant frictional torque of the bearings. Since ω = constant, $T_a = T_{\text{bearings}}$. Due to wear in the drive system, the applied torque develops an oscillating component so that $T_a(t) = T_a + T_o \sin \omega t$. Add a shaft and disk to the system that absorbs the oscillating torque so that the original disk does not oscillate. Support your design with an analysis that shows how the oscillation can be eliminated.

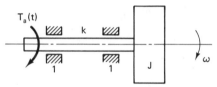

5.9. The shaft shown in the diagram is subjected to an oscillating torque $(T_o \cos \omega t)$ applied at disk J_2. Use Lagrangian analysis to find the equations of motion, and then determine the steady-state amplitudes of θ_1 and θ_2. Neglect the transient response.

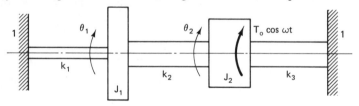

5.10. The diagram shows two disks (J_1 and J_2) and supporting massless shafts (k_1 and k_2).
$J_1 = 50$ lb$_f$-in.-s^2
$J_2 = 100$ lb$_f$-in.-s^2
$k_1 = 10^5$ lb$_f$-in./rad
$k_2 = 4(10^5)$ lb$_f$-in./rad

For this rotational system, determine the following:
(a) The equations of motion.
(b) The characteristic equation.
(c) The *exact* natural frequencies.

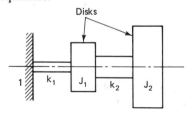

5.11. Determine the equation for the disturbing torque absorbed by the pendulum absorber for a small-amplitude shaft disturbance and a small-amplitude displacement of the absorber arm. Include the inertia of the absorber's disk of radius R.

5.12. Section 5.10 refers to a third Lagrange equation, which was not used in the analysis presented there.
   (a) Identify that equation.
   (b) Obtain the differential equation that it yields.
   (c) Simplify the differential equation.
   (d) Draw a (d'Alembert) free-body diagram for the disk and verify that your answer is in agreement.

5.13. The diagram shows two disks mounted on a lightweight shaft [$EI = 6(10^{-3})$ lb$_f$-in.^{2}]. Using the Dunkerley method, determine the lowest natural frequency.

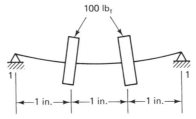

5.14. Disks I_1 and I_2 are joined by a torsional shaft (spring k_2) and to ground by a torsional shaft (spring k_1).
   $k_1 = k_2 = 1.0$ lb$_f$-ft/rad
   $I_1 = I_2 = 1.0$ lb$_f$-ft-s^2
   (a) Find the flexibility matrix for this system.
   (b) Using the flexibility matrix, find the natural frequencies of the system.

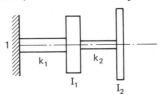

5.15. The diagram shows a cantilevered disk bent by dynamic forces.
   (a) If the shaft $\dot{\phi}$ is constant and the disk is unbalanced ($e \neq 0$), determine the effect that the shaft motion has on the shaft's lateral force F and bending couple M. Assume that the center of mass g hangs outside, as shown, and always remains in the plane of δ (analogous to synchronous whirl below the first bending critical).
   (b) Let the unbalance $e = 0$. Also let ϕ and θ vary with time, as would be expected during a speed change. Show how $\ddot{\phi}$ and $\dot{\phi}$, $\ddot{\theta}$, $\dot{\theta}$, etc. are related to the shaft's input driving torque T_s.

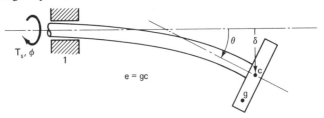

5.16. The diagram shows a rotating coordinate system that rotates with the shaft, which has different stiffnesses in two orthogonal directions. This coordinate system is aligned with the shaft's cross-section principal axes. Using these coordinates, determine the following:

(a) Obtain the expression for the kinetic energy of the disk when ω = constant.

(b) Obtain the expression for the potential energy of the shaft when ω = constant, neglecting gravity effects.

(c) Using the Lagrange equations, obtain the differential equations of motion for x_c and y_c.

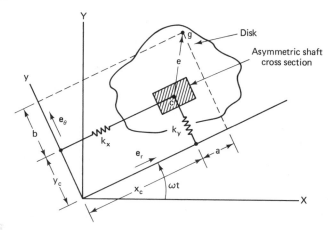

5.17. The diagram shows an end view of a disk symmetrically mounted on a flexible shaft (k_r). The bearing B is supported on an unequal-stiffness base with stiffnesses k_x and k_y that permit the bearing centers B to displace from their rest position O. Derive the spring force components acting at c in terms of x_c, y_c, k_x, k_y, and k_r.

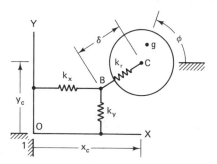

5.18. The diagram shows a single symmetrically mounted disk on an elastic shaft with stiffness k_r. The shaft supports at the ends are also elastic, with different total horizontal and vertical stiffnesses as shown. The shaft rotates at a constant angular velocity ω.

(a) Write the potential energy of the system in terms of the quantities x_B, y_B, δ, e, ϕ, θ, k_x, k_y, and k_r.

(b) Obtain the horizontal and vertical velocity components in terms of the quantities shown in their derivatives.

(c) Obtain an alternate expression for the potential energy in terms of x_c, y_c, k_x, k_y, and k_r, where x_c and y_c are the coordinates of disk center c.

End view

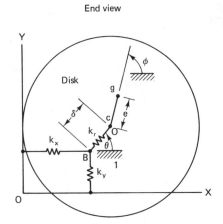

5.19. The diagram shows a 1500-lb shaft supported by flexible bearings that give cantile-vered end constraints. Simple rigid (knife-edge) supports are also shown. Real bear-ings lie somewhere between these extremes. Assuming that the base supporting the bearings has a $k_x = 1.0(10^6)$ lb$_f$/in. and $k_y = 1.5(10^6)$ lb$_f$/in. at each end, determine the range of shaft rpm's in which the critical speed would be expected to lie.

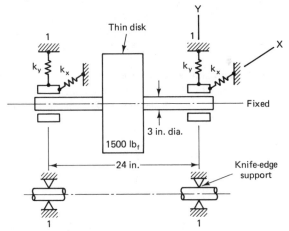

5.20. The bearing supports at each end of the shaft shown in the diagram have flexibility equivalent to a spring constant of $2.5(10^5)$ lb$_f$/in. perpendicular to the shaft in any di-rection. Due to bending alone, the shaft has a displacement of 0.0018 inches at the 300-lb$_f$ disk. What effect does the bearing flexibility have on the system critical speed?

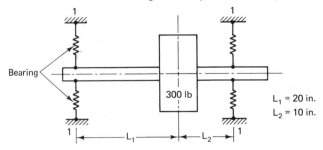

5.21. Your company decides to change the dimensions and materials used to make the armature–shaft assembly shown in the diagram. They are going to halve the bearing span L, halve the assembly weight W, and halve the modulus E. Having done this, they do not want the natural frequency to change, i.e., it is to remain at its original value. Your job is to determine what the new shaft diameter should be relative to the old value so that the natural frequency of the new design equals the natural frequency of the old design.

The displacement under load for a centrally loaded beam with simple supports is

$$\delta_{max} = \frac{WL^3}{48EI} \qquad I_{shaft} = \frac{\pi d^4}{64}$$

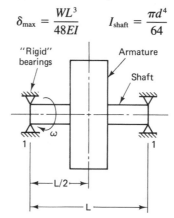

5.22 A 100-kg rotor is to operate at 2000 rpm. The spring constant for the symmetric shaft (in rigid bearings and supports) is 10^4 N/cm. The unbalance is $e = 0.010$ cm.
 (a) What is the shaft critical speed?
 (b) What is the shaft displacement at the running speed?
 (c) What is the force exerted on the bearings by the shaft imbalance?

5.23. The diagram shows a mass–spring system with frictionless rollers to guide the masses so they can only move vertically.
 (a) Obtain an estimate of the lowest natural frequency of the system using the Rayleigh method.
 (b) Compare this result with the exact natural frequencies.

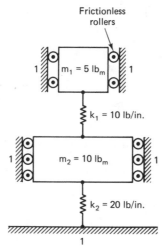

5.24. The diagram shows a crude model of a two-story building, where the floors are lumped as m_1 and m_2 and the supporting walls are lumped as displacement (but not torsional) springs. Assuming that the masses only move horizontally (as shown) and do not rotate, obtain the characteristic equation for the system.

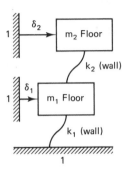

5.25. A 100-lb-weight disk is placed on a lightweight beam, as shown in the diagram. Strength-of-materials texts show that the beam loadings shown (F and M) yield linear and angular displacements *at the disk* location given by the flexibility matrix formulation:

$$\begin{Bmatrix} y \\ \theta \end{Bmatrix} = \frac{1}{EI}\begin{bmatrix} 4/9 & -2/9 \\ -2/9 & 2/3 \end{bmatrix}\begin{Bmatrix} F \\ M \end{Bmatrix}$$

Assuming that the y and θ are harmonic, obtain the characteristic equation for the natural frequencies of the system. Do not solve this equation for the natural frequencies. The beam does not rotate!

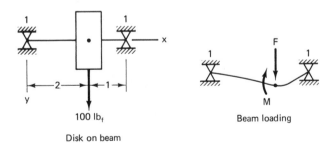

Disk on beam   Beam loading

5.26. Mass m is mounted on a vertical cantilevered beam. Neglecting gravity effects, find the characteristic equation assuming that θ and δ are harmonic motions. The beam length is l. The moment of inertia of the mass about g is J_g. Consider the beam to be massless.

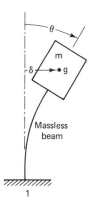

5.27. The diagram shows a device that can measure the angular velocity of the massless platform to which it is attached. It does so by turning through angle θ relative to the platform as the platform spins about the vertical with $\dot{\phi}$ = constant. Two equal springs (k) restrain the angle θ to very small values. The disk speed about its own axis is $\dot{\psi}$ = constant.

(a) Obtain the kinetic energy T for this system.

(b) Obtain the potential energy V for this system.

(c) Using the appropriate Lagrange equation, obtain the differential equation of motion.

(d) Find the natural frequency of the system.

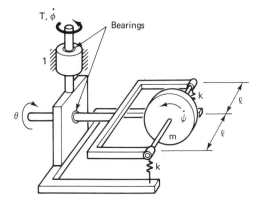

5.28 The diagram shows a steel disk and shaft supported on ball bearings. The shaft speed is 800 rpm. The design is considered inadequate if the first critical speed is within 20% of the running speed. Is the design adequate? $k_x = k_y$ = bearing stiffness. (Correct for gyroscopic effect.)

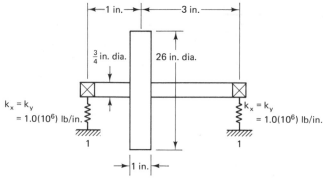

Steel sp. ωt. = 490 lb$_f$/ft^3

5.29. The diagram shows a rubber-tired 14-in. wheel attached to a trailer by an axle. An *xyz* coordinate system with its origin is also shown. The coordinates of the center of mass of the 50-lb$_f$ wheel–tire assembly are $x = 0.1$ in., $y = 0.15$ in., and $z = 3.2$ in. in this coordinate system. The wheel turns at 600 rpm.
 (a) Before the assembly is balanced, determine the force and moment exerted at O by the imbalance of the assembly.
 (b) Determine the size of the balance weights that must be attached to the wheel rim to obtain a dynamic balance.

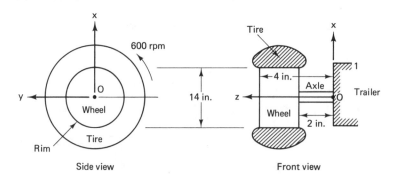

Side view   Front view

5.30. The diagrams show a rigid shaft with a disk at midspan mounted on a balancing machine. The ends of the shaft pass through frictionless bearings in blocks that are constrained to move horizontally by guideways. These blocks, each weighing 1.0 lb$_f$, are restrained by springs, each with a $k = 1000$ lb$_f$/in. The disk and shaft weight 32 lb$_f$. The shaft is driven at a constant $\omega = 60$ rad/s. Sensors measure each block's horizontal displacement. The peak-to-peak amplitude measured by them is 0.170 inches. The blocks slide along the guideways without friction. When the blocks are centered as shown in the side view, the springs exert no force.
 (a) Obtain the kinetic energy for the system.
 (b) Obtain the potential energy for the system, neglecting the effects of gravity.
 (c) Using the appropriate Lagrange equation, obtain the differential equation for the horizontal displacement of the blocks.
 (d) Solve the differential equation for the steady-state displacement of the blocks. As-

sume that the small amount of damping present causes the starting transient to vanish.

(e) Calculate the unbalance distance *e* in inches.

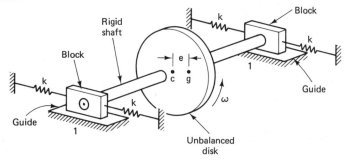

Schematic

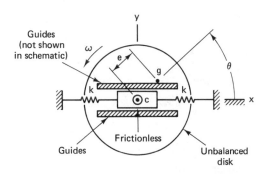

Side view

5.31 The diagram shows an unbalanced disk (with a short shaft) mounted on a balancing machine. When the disk is not rotating, the gap-sensing device is "zeroed" to read zero. When rotated at a constant speed, the sensor shows that the gap is reduced a maximum of 0.020 inches at the 0° point on the disk. A 0.01-lb$_f$ trial weight is added to the rim at 90°. The maximum reduction in gap is now 0.040 inches at 45°. What size balance weight should be added to the rim and what is its angular location?

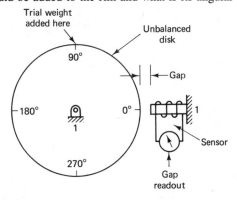

5.32. Using Equations (5.57) obtain formulas for the following:
   (a) m_b/m_t
   (b) angle ϕ.
   These formulas should not contain $i = \sqrt{-1}$

5.33. The diagram shows a cylindrical rotor unbalanced by small masses m_1 and m_2. The rotor is mounted in a balancing machine with frictionless bearings. The coordinate system shown has its z coordinate attached to the cylinder centerline. The y coordinate is always vertical and the x coordinate is always horizontal.
   (a) Write the Lagrangian for this system, assuming that the yaw angle ψ of the z axis remains very small.
   (b) Obtain the differential equations for X and ψ. Identify the cross-coupling coefficients.

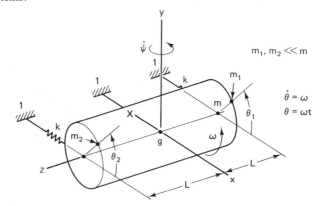

INDEX